The Melanotropic Peptides

Volume III: Mechanisms of Action and Biomedical Applications

Editor

Mac E. Hadley, Ph.D.
Department of Anatomy
University of Arizona
Tucson, Arizona

CRC Press
Taylor & Francis Group
Boca Raton London New York

CRC Press is an imprint of the
Taylor & Francis Group, an **informa** business

First published 1988 by CRC Press
Taylor & Francis Group
6000 Broken Sound Parkway NW, Suite 300
Boca Raton, FL 33487-2742

Reissued 2018 by CRC Press

Library of Congress Cataloging-in-Publication Data

The Melanotropic peptides / editor. Mac E. Hadley.
p. cm.
Contents: v. 1. source, synthesis, chemistry, secretion, circulation, and metabolism — v. 2. Biological roles — v. 3. Mechanisms of action and biomedical applications.
Includes bibliographies and index.
1. MSH (Hormone) I. Hadley, Mac E.
QP572.M75M45 1988
[DNLM: 1. MSH--physiology. WK 515 M517]
596'.0142--dc19
DNLM/DLC
ISBN 0-8493-5277-0 (v. 1)
ISBN 0-8493-5278-9 (v. 2) 87-29903
ISBN 0-8493-5279-7 (v. 3)

ISBN 13: 978-1-315-89528-4 (hbk)
ISBN 13: 978-1-351-07438-4 (ebk)

Visit the Taylor & Francis Web site at http://www.taylorandfrancis.com and the
CRC Press Web site at http://www.crcpress.com

PREFACE

The three volumes on "*The Melanotropic Peptides*" are the outcome of a conference of the same name that was held in Tucson, Arizona, from October 11-12, 1986.

The melanotropic peptides are the hormones that are generally recognized as playing a major role in the control of vertebrate integumental pigmentation. The two physiologically relevant melanotropins are α-melanocyte-stimulating hormone (α-MSH, α-melanotropin) and melanin-concentrating hormone (MCH). Because α-MSH and corticotropin (adrenal cortical-stimulating hormone, ACTH) share some structural similarity, these other melanocorticotropins possess MSH-like activity when used at higher (unphysiological) concentrations. A quick glance at the topics that are included within these volumes makes it clear that these peptidergic hormones probably also play important roles in the control of a number of diverse physiological processes.

The format of the three volumes provides a complete coverage of what is known about the Melanotropic Peptides. Volume I provides information on the source, synthesis, chemistry, mechanism of secretion, control of secretion, and the circulation and metabolism of the melanotropic peptides. In addition, a chapter on the pharmacokinetics of secreted MSH as determined by RIA, is provided.

Volume II discusses the diverse putative physiological roles of MSH and MCH. A concluding chapter of this volume provides a discussion of the possible evolutionary relationships between MSH and MCH in the control of color change. Volume III discusses the known mechanisms of action of the two melanotropic peptides, and concludes with a discussion of the possible biomedical applications of the melanotropins.

As editor, I am indebted to each of the authors who have contributed their time, effort, and expertise to their particular chapter. Although we can, of course, expect advancements in our knowledge of the melanotropins, these volumes will provide the primary source of information on the melanotropic peptides for many years to come.

THE EDITOR

Mac Eugene Hadley, Ph.D., is a professor in the Department of Anatomy at the University of Arizona. He also holds a joint appointment with the Department of Molecular and Cellular Biology.

Dr. Hadley received his doctorate degree from Brown University in 1966. He then did his postdoctoral work at the University of Arizona after which he became an assistant professor, associate professor, and a full professor at the same institution.

Dr. Hadley was trained as a biologist and developed a particular interest in endocrinology. He has written a popular textbook of *Endocrinology* which is used in many college classes.

Dr. Hadley's major research interest has been the endocrine control of vertebrate integumental pigmentation. He has published numerous manuscripts and reviews in leading scientific journals. He is a member of the International Pigment Cell Society which presented him the Myron Gordon Award in 1986 for his research in the area of the neuroendocrine control of pigmentation.

The broad scope of Dr. Hadley's research interests is reflected in the range of topics that comprise the three volumes on "*The Melanotropic Peptides.*"

CONTRIBUTORS

Volume I

Saida Adjeroud, Ph.D.
INSERM
Faculté des Sciences
Université de Rouen
Mort-St.-Aignan, France

Ann C. Andersen, Ph.D.
INSERM
Faculté des Sciences
Université de Rouen
Mont-St.-Aignan, France

Arlette Burlet, Ph.D.
Laboratoire d'Histologie A
INSERM
Faculté de Medecine
Vandoeuvre-les-Nancy
France

Ana Maria de Lauro Castrucci, Ph.D.
Assistant Professor
Department of Physiology
University of São Paulo
São Paulo, Brazil

Bibie M. Chronwall, Ph.D.
Associate Professor
University of Missouri-Kansas City
School of Basic Life Sciences
Division of Structural and Systems Biology
Kansas City, Missouri

Jean-Michel Danger, Ph.D.
INSERM
Faculté des Sciences
Université de Rouen
Mont-St.-Aignan, France

Robert M. Dores, Ph.D.
Assistant Professor
Department of Biological Sciences
University of Denver
Denver, Colorado

Stela Elkabes, Ph.D.
Visiting Fellow
National Institutes of Health
Bethesda, Maryland

Mac E. Hadley, Ph.D.
Professor
Department of Anatomy
University of Arizona
Tucson, Arizona

Victor J. Hruby, Ph.D.
Professor
Department of Chemistry
University of Arizona
Tucson, Arizona

Bruce G. Jenks, Ph.D.
Department of Animal Physiology
University of Nijmegen
Nijmegen, The Netherlands

Hiroshi Kawauchi, Ph.D.
Professor
Laboratory of Molecular Endocrinology
School of Fisheries Sciences
Kitasato University
Sanriku, Iwate, Japan

Gotfryd Kupryszewski, Ph.D.
Laboratory of Bioorganic Chemistry
Institute of Chemistry
University of Gdansk
Gdansk, Poland

Marek Lamacz, Ph.D.
INSERM
Faculté des Sciences
Université de Rouen
Mont-St.-Aignan, France

Philippe Leroux, Ph.D.
INSERM
Faculté des Sciences
Université de Rouen
Mont-St.-Aignan, France

Y. Peng Loh, Ph.D.
National Institutes of Health
Bethesda, Maryland

Philip J. Lowry, Ph.D., D.Sc., F.I. Biol.
Professor
Department of Physiology and Biochemistry
University of Reading
Reading, England

Gerard J. M. Martens, Ph.D.
Research Associate
Departments of Animal Physiology and Molecular Biology
University of Nijmegen
Nijmegen, The Netherlands

Brenda Myers, B.A.
Student
Department of Biology
Johns Hopkins University
Baltimore, Maryland

Tom O'Donohue, Ph.D. (Deceased)
Director
CNS Research
Monsanto Company
St. Louis, Missouri

Georges Pelletier
MRC Group for Molecular Endocrinology
Centre Hospitalier de L'Université Laval
Laval, Quebec, Canada

Elizabeth K. Perryman, Ph.D.
Professor
Department of Biological Sciences
California Polytechnic State University
San Luis Obispo, California

Lise Stoeckel, Ph.D.
CNRS
Laboratoire de Physiologie Générale
Université Louis Pasteur
Strasbourg, France

Marie-Christine Tonon, Ph.D.
INSERM
Faculté des Sciences
Université de Rouen
Mont-St.-Aignan, France

Hubert Vaudry, D.Sc.
Directeur de Recherche
INSERM
Départment d'Endocrinologie Moleculaire
Université de Rouen
Mont-St.-Aignan, France

B. M. Lidy Verburg-van Kemenade, Ph.D.
Department of Animal Physiology
University of Nijmegen
Nijmegen, The Netherlands

John F. Wilson, Ph.D.
Department of Pharmacology and Therapeutics
University of Wales College of Medicine
Cardiff, Wales

CONTRIBUTORS

Volume II

Zalfa A. Abdel-Malek, Ph.D.
Assistant Professor
Department of Dermatology
College of Medicine
University of Cincinnati
Cincinnati, Ohio

Joseph T. Bagnara, Ph.D.
Professor
Department of Anatomy
University of Arizona
Tucson, Arizona

Bridget I. Baker, Ph.D.
Doctor
School of Biological Sciences
University of Bath
Bath, England

C. Wayne Bardin, M.D.
Director of Biomedical Research
Center for Biomedical Research
The Population Council
New York, New York

Bill E. Beckwith, Ph.D.
Associate Professor
Department of Psychology
University of North Dakota
Grand Forks, North Dakota

Carla Boitani, Ph.D.
Research Associate
Institute of Histology and General Embryology
University of Rome, "La Sapienza"
Rome, Italy

Ana Maria de Lauro Castrucci, Ph.D.
Assistant Professor
Department of Physiology
University of São Paulo
São Paulo, Brazil

Ching-Ling Chen, Ph.D.
Scientist
Center for Biomedical Research
The Population Council
New York, New York

P. De Koning, Ph.D.
Institute of Molecular Biology and Medical Biotechnology
University of Utrecht
Utrecht, The Netherlands

A. Dell, Ph.D.
Department of Biochemistry
Imperial College London
London, England

P. M. Edwards, Ph.D.
Institute of Molecular Biology and Medical Biotechnology
University of Utrecht
Utrecht, The Netherlands

W. H. Gispen, Ph.D.
Professor
Institute of Molecular Biology and Medical Biotechnology
University of Utrecht
Utrecht, The Netherlands

Mac E. Hadley, Ph.D.
Professor
Department of Anatomy
University of Arizona
Tucson, Arizona

K. L. Henville, Ph.D.
Department of Biochemistry
St. Bartholomew's Hospital Medical College
London, England

J. P. Hinson, Ph.D.
Department of Biochemistry
St. Bartholomew's Hospital Medical College
London, England

Omid Khorram, Ph.D.
Medical Student IV
School of Medicine
Texas Tech University
Lubbock, Texas

James M. Lipton, Ph.D.
Professor
Department of Physiology
University of Texas
Health Science Center at Dallas
Dallas, Texas

Samuel M. McCann, Ph.D.
Professor
Department of Physiology
University of Texas
Health Science Center at Dallas
Dallas,Texas

H. R. Morris, Ph.D.
Department of Biochemistry
Imperial College London
London, England

Patricia L. Morris, Ph.D.
Research Investigator
Center for Biomedical Research
The Population Council
New York, New York

Chandrima Shaha, Ph.D.
National Institute of Immunology
JNU Complex
New Delhi, India

Wade C. Sherbrooke, Ph.D.
Resident Director
Southwestern Research Station
American Museum of Natural History
Portal, Arizona

Anthony J. Thody, Ph.D.
Reader in Experimental Dermatology
Department of Dermatology
University of Newcastle-upon-Tyne
Newcastle-upon-Tyne, England

C. E. E. M. Van der Zee, Ph.D.
Institute of Molecular Biology and Medical Biotechnology
University of Utrecht
Utrecht, The Netherlands

R. Gerritsen van der Hoop, Ph.D.
Institute of Molecular Biology and Medical Biotechnology
University of Utrecht
Utrecht, The Netherlands

J. Verhaagen, Ph.D.
Institute of Molecular Biology and Medical Biotechnology
University of Utrecht
Utrecht, The Netherlands

Gavin P. Vinson, D.Sc.
Professor
Department of Biochemistry
St. Bartholomew's Hospital Medical College
London, England

Brian Weatherhead, M.A., Ph.D., C.Biol., F.I. Biol.
Professor and Head of Department
Department of Anatomy
University of Hong Kong
Hong Kong

B. J. Whitehouse, Ph.D.
Department of Physiology
King's College London (KQC)
London, England

Catherine A. Wilson, Ph.D.
Doctor
Department of Obstetrics and Gynaecology
St. George's Hospital Medical School
London, England

CONTRIBUTORS

Volume III

Fahad Al-Obeidi, Ph.D.
Department of Chemistry
University of Arizona
Tucson, Arizona

Ana Maria de Lauro Castrucci, Ph.D.
Assistant Professor
Department of Physiology
University of São Paulo
São Paulo, Brazil

Dhirendra Chaturvedi, Ph.D.
Senior Chemist
Research and Development Laboratories
Vega Biotechnologies, Inc.
Tucson, Arizona

Wayne L. Cody, Ph.D.
Research Laboratory
Eastman Chemicals Company
Kingsport, Tennessee

Pierre N. E. de Graan, Ph.D.
Doctor
Department of Molecular Neurobiology
Rudolf Magnus Institute for Pharmacology and
Institute of Molecular Biology and Biotechnology
University of Utrecht
Utrecht, The Netherlands

Alex N. Eberle, Ph.D.
Head, Laboratory of Endocrinology
University Hospital
Basel, Switzerland

Jürg Girard, Ph.D.
Professor
Department of Pediatric Endocrinology
University Hospital
Basel, Switzerland

Mac E. Hadley, Ph.D.
Professor
Department of Anatomy
University of Arizona
Tucson, Arizona

Victor J. Hruby, Ph.D.
Professor
Department of Chemistry
University of Arizona
Tucson, Arizona

Aaron B. Lerner, Ph.D.
Department of Dermatology
Yale University School of Medicine
New Haven, Connecticut

Gisela E. Moellmann, Ph.D.
Associate Professor
Department of Dermatology
Yale University School of Medicine
New Haven, Connecticut

John M. Pawelek, Ph.D.
Senior Research Scientist
Department of Dermatology
Yale University School of Medicine
New Haven, Connecticut

Tomi K. Sawyer, Ph.D.
Senior Research Scientist III
Biotechnology Division
The Upjohn Company
Kalamazoo, Michigan

Walter Siegrist, Ph.D.
Professor
Department of Pediatric Endocrinology
University Hospital
Basel, Switzerland

J. W. Sam Stevenson, Ph.D.
Research Chemist
Department of Physical and Analytical Chemistry Research
Tennessee Eastman Company
Kingsport, Tennessee

Elizabeth Sugg, Ph.D.
Department of Chemistry
University of Arizona
Tucson, Arizona

TABLE OF CONTENTS

Volume I

Chapter 1
The Melanotropic Peptides: An Introduction 1
Philip J. Lowry

Chapter 2
Melanotropins: Pars Intermedia Structure and Secretion 5
Elizabeth K. Perryman

Chapter 3
Pituitary Melanotropin Biosynthesis 25
Robert M. Dores

Chapter 4
The Melanotropic Peptides: Structure and Chemistry 39
Hiroshi Kawauchi

Chapter 5
Melanocyte-Stimulating Hormone in the Central Nervous System 55
Bibie Chronwall and Thomas L. O'Donohue

Chapter 6
The Pro-Opiomelanocortin Gene in *Xenopus Laevis:* Structure, Expression, and Evolutionary Aspects 67
Gerard J. M. Martens

Chapter 7
Regulation of Pro-Opiomelanocortin Biosynthesis in the Amphibian and Mouse Pituitary Intermediate Lobe 85
Y. Peng Loh, Stella Elkabes, and Brenda Myers

Chapter 8
Pro-Opiomelanocortin in the Amphibian Pars Intermedia: a Neuroendocrine Model System 103
Bruce G. Jenks, B. M. L. Verburg-van Kemenade, and Gerard J. M. Martens

Chapter 9
Multihormonal Control of Melanotropin Secretion in Cold-Blooded Vertebrates 127
M. C. Tonon, J. M. Danger, M. Lamacz, P. Leroux, S. Adjeroud, A. Andersen, B.M.L. Verburg-van Kemenade, B. G. Jenks, G. Pelletier, L. Stoekel, A. Burlet, G. Kupryszewski, and H. Vaudry

Chapter 10
Melanotropin Enzymology 171
Ana Maria de L. Castrucci, Mac E. Hadley, and Victor J. Hruby

Chapter 11
Peripheral and Central Pharmacokinetics of the Melanotropins 183
John F. Wilson

Index 211

TABLE OF CONTENTS

Volume II

Chapter 1
Melanotropins and Melanin Pigmentation of the Skin of Mammals 1
Brian Weatherhead

Chapter 2
Melanotropins, Chromatophores, and Color Change 21
Joseph T. Bagnara

Chapter 3
Melanotropin Effects on Pigment Cell Proliferation 29
Zalfa A. Abdel-Malek

Chapter 4
The Melanotropins, Learning and Memory 43
Bill E. Beckwith

Chapter 5
Melanotropins and Peripheral Nerve Regeneration 73
C. E. E. M. Van der Zee, P. M. Edwards, R. Gerritsen van der Hoop, P. De Koning, J. Verhaagen, and W. H. Gispen

Chapter 6
The Actions of α-MSH on the Adrenal Cortex 87
Gavin P. Vinson, B. J. Whitehouse, K. L. Henville, J. P. Hinson, A. Dell, and H. R. Morris

Chapter 7
MSH in CNS Control of Fever and its Influence on Inflammation/Immune Responses .. 97
James M. Lipton

Chapter 8
The Presence and Possible Function of α-MSH and Other POMC-Derived Peptides in the Reproductive Tract 115
Carla Boitani, Patricia L. Morris, Ching-Ling Chen, Chandrima Shaha, and C. Wayne Bardin

Chapter 9
The Role of Melanotropins in Sexual Behavior 131
Anthony J. Thody and Catherine A. Wilson

Chapter 10
Neuroendocrine Effects of Melanotropins 145
Omid Khorram and Samuel M. McCann

Chapter 11
Melanin-Concentrating Hormone 159
Bridget I. Baker

Chapter 12
Melanotropic Peptides and Receptors: An Evolutionary Perspective in Vertebrates..... 175
Wade C. Sherbrooke, Mac E. Hadley, and Ana Maria de L. Castrucci

Index 191

TABLE OF CONTENTS

Volume III

Chapter 1
Melanotropin Bioassays 1
Mac E. Hadley and Ana Maria de L. Castrucci

Chapter 2
Melanotropin Mechanisms of Action: Melanosome Movements 15
Mac E. Hadley and Ana Maria de L. Castrucci

Chapter 3
The Role of Protein Phosphorylation in the Mechanism of Action of α-MSH 27
Pierre N. E. De Graan and Alex N. Eberle

Chapter 4
Melanotropin Mechanisms of Action: Melanogenesis 47
John M. Pawelek, John McLane, and Michael Osber

Chapter 5
Melanotropins and Receptor Signal Transduction 59
Tomi K. Sawyer, Mac E. Hadley, and Victor J. Hruby

Chapter 6
Cyclic Conformationally Constrained Melanotropin Analogues: Structure-Function and Conformational Relationships 75
Wayne L. Cody, Mac E. Hadley, and Victor J. Hruby

Chapter 7
Melanotropin Three-Dimensional Structural Studies by Physical Methods and Computer-Assisted Molecular Modelling ....
Wayne L. Cody, J. W. Sam Stevenson, Fahad Al-Obeidi, Elizabeth Sugg, and Victor J. Hruby

Chapter 8
Melanotropin Receptors: Studies with Labeled Melanotropins 111
Alex N. Eberle, Pierre N. E. De Graan, Walter Siegrist, and Jürg Girard

Chapter 9
Melanotropins: Biomedical Applications 129
Dhirendra Chaturvedi and Mac E. Hadley

Chapter 10
The Melanotropic Peptides: A Summary 151
Aaron B. Lerner and Gisela E. Moellmann

Index 153

Chapter 1

MELANOTROPIN BIOASSAYS

Mac E. Hadley and Ana Maria de L. Castrucci

TABLE OF CONTENTS

I. Introduction 2
A. The Melanophore Index (MI) 2

II. Frog Skin Bioassays 2
A. In Vitro Bioassay 2
B. In Vivo Bioassay 4

III. Lizard (*Anolis*) Skin Bioassay 4

IV. Teleost (Fish) Skin Bioassay 6

V. Melanoma Bioassays 7
A. Melanoma Tyrosinase Bioassay 8
B. Melanoma Adenylate Cyclase Bioassay 10
C. Melanoma Clonogenic Bioassay 11
D. *In Situ* Melanin Assay 11

VI. Prolonged (Residual) Actions of Melanotropins 12

VII. Perspectives 12

Acknowledgments 12

References 12

I. INTRODUCTION

Bioassays have played an important role in determining the presence of hormones in the blood or other tissues as well as determining the biological potency of structural analogs of hormones. The frog skin bioassay, developed by Shizume et al. in 1954, has been particularly important in advancing our knowledge about the melanotropins. This bioassay, like the earlier assays for melanotropins, involved observations or recordings of the responses of pigment cells (chromatophores, particularly melanophores) present within the skin. Both in vitro as well as in vivo melanotropin bioassays have been employed in several vertebrate species. Melanoma cells provide a unique bioassay for determining the biochemical correlates of melanotropin action. In the present report some of the more commonly utilized bioassays for melanotropic peptides will be described. For a detailed discussion of the early bioassay systems for the melanotropins, several references are available,[2-6] in addition to the classical monograph by Parker.[7] For information on the radioimmunoassay of melanotropins several references are now available (see Wilson, Volume I, Chapter 11).

A. The Melanophore Index (MI)

The earliest studies of melanophore responses to melanotropins utilized a so-called "melanophore index" (MI).[8] By this method the response of melanophores to a variety of stimuli could be subjectively quantitated by assigning numbers to the observed degree of melanosome (melanin granule) dispersion within the cells. In response to melanotropins, the perinuclearly aggregated (concentrated) melanosomes (MI-1) begin to disperse out into the dendritic processes of the melanophore until a maximum response (MI-5) is obtained (Figure 1). This assay is presently still used by many investigators throughout the world to study melanophore responses of teleosts, amphibians, and reptiles to melanotropic peptides. An analogous "chromatophore index" can be similarly utilized to study the response of other integumental chromatophore types (iridophores, erythrophores, and xanthophores) to melanotropins.

II. FROG SKIN BIOASSAYS

The skin of the frog has been most widely used for the study of the melanotropic activity of peptides. Both in vitro and in vivo bioassays have been utilized. Both assays involve the response of integumental chromatophores to stimulation by melanotropins. This response to exogenous melanotropins is identical to that resulting from the release of endogenous melanotropins from the pars intermedia of the pituitary gland. Melanotropins stimulate the dispersion of melanosomes out into the dendritic processes of melanophores and concomitantly stimulate the aggregation of reflecting platelets within iridophores. The composite structural unit of color change is referred to as the "chromatophore unit" (see Bagnara, Volume II, Chapter 2).

A. In Vitro Frog Skin Bioassay

Skins from the leopard frog, *Rana pipiens*, have been most commonly employed in the in vitro bioassay. However, skin from other ranid species as well as toads, appears to be equally useful. The bullfrog, *R. catesbeiana,* has a greater worldwide availability and, like *R. pipiens*, is very sensitive to melanotropins. In the present report, the details provided are equally applicable to both species of frog. The real advantage of *R. catesbeiana* is that the frogs are generally larger than other ranids, and most of the dorsal body skin can also be utilized in addition to skin from the legs.

Frogs are sacrificed by decapitation and/or pithing, and the leg and the thigh skins are removed from the animals and placed in a physiological saline (Ringer) solution. The skins are then mounted on metal rings and held in place by an outer plastic ring as described for

FIGURE 1. Melanophore index (MI). Melanotropins stimulate melanosome dispersion; melanosomes migrate from a perinuclear position within the melanophore (MI-1) out into the dendritic processes of the pigment cell (MI-5).

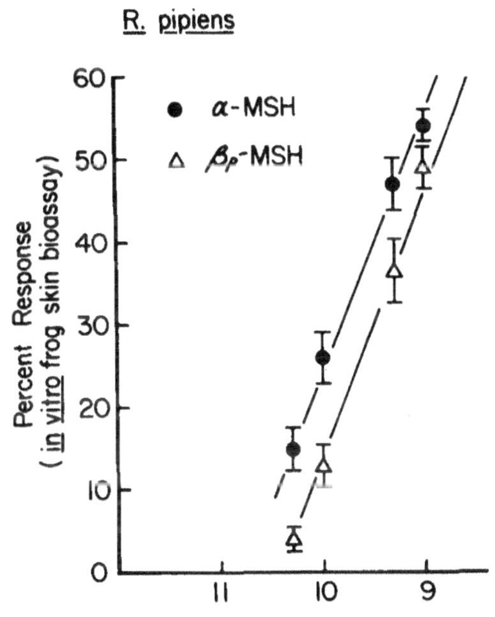

FIGURE 2. In vitro frog (*R. pipiens*) skin bioassay. Four-point dose-response curves to α- and β-MSH. All points are significantly different ($p < 0.05$) from the control set of skins. (n = 16 per point).

the original bioassay.[1] The mounted skins are then placed in beakers containing Ringer solution. The addition of a melanotropin to the bathing solution results in darkening of the skins, and subsequent removal of the peptide by placing the skins in Ringer in the absence of the peptide results in a relightening of the skins. Changes in skin color (reflectance) are monitored by a Photovolt reflectometer and recorded as percent changes from the initial base (zero) value. In three species of ranids studied α-melanocyte-stimulating hormone (α-MSH) was found to be slightly (2—4 fold) more potent than β-MSH.[2] The minimal effective dose found to significantly darken the skins above control levels is about 2 to 4 $\times 10^{-11}$ *M* (Figure 2). A maximal response is obtained at about 60 min following addition of a melanotropin, and the skins are light again by 90 to 120 min following removal of the peptides (Figure 3). Therefore, skins can be used several times in one day. All melanotropin analogs studied to date have provided parallel dose response curves (see Cody et al.,[9] Chapter 7, this Volume).

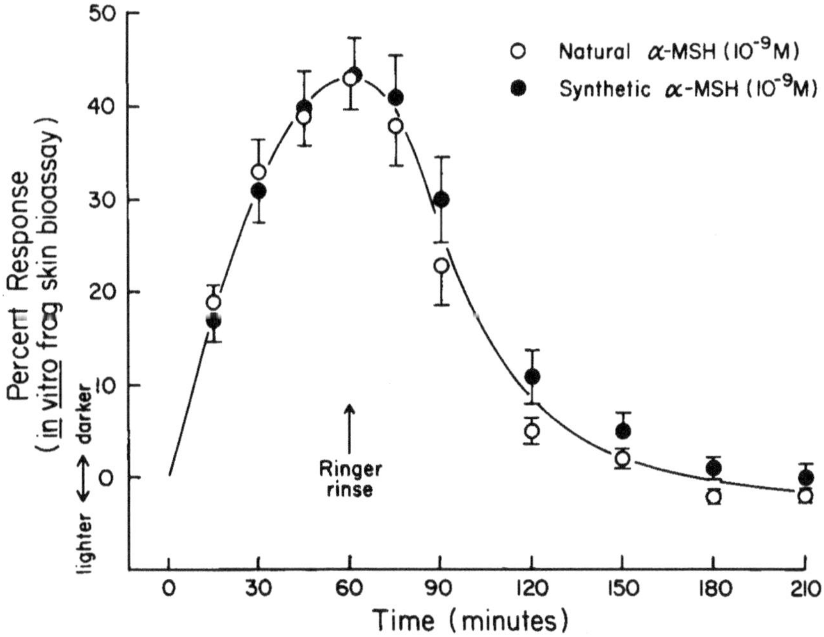

FIGURE 3. Reversal of the darkening response of frog skins to natural and synthetic α-MSH, after transfer of the skins to Ringer in the absence of the melanotropins (n = 6).

B. In Vivo Frog Skin Bioassay

We recently described an in vivo bioassay for melanotropins using the frog, *R. pipiens.*[2] The reflectance method was again used, and we determined the in vivo potencies of α- and β-MSH (Figure 4A and B). Unexpectedly, the melanotropins were as active in the in vivo assay as in the in vitro bioassay. The frogs darkened in a dose-response manner to the injected melanotropins. α-MSH as low as 5×10^{-11} mmol/10 g/frog induced significant darkening of the animals. Both α- and β-MSH, but particularly α-MSH, are rapidly inactivated in frog serum in vitro (see Castrucci and Hadley, Chapter 1, this volume). Therefore, it was surprising that an injection of about 5×10^{-10} mmol/10 g body weight of α-MSH gave a response approximately equal to 10^{-9} mmol of peptide/10 mℓ of bathing medium in vitro.

Kastin et al.[10] used hypophysectomized *R. pipiens* and found by using the classical MI that the in vivo assay was as sensitive as the in vitro assay for the determination of the potency of a number of adrenocorticotropin (ACTH)-like peptides. These observations are even more surprising since these peptides unlike α-MSH, which is N-terminal acetylated and C-terminal amidated, should be even more vulnerable to inactivation by exopeptidases. In this regard, one might even expect that β-MSH, although somewhat longer than α-MSH, would be more susceptible than the shorter α-MSH to inactivation by exopeptidases. The actions of β-MSH in vivo are more prolonged than an equimolar concentration of α-MSH when injected into the frog (Figure 4 A and B). The more prolonged actions of β-MSH may relate to the fact that the peptide is less susceptible to inactivation by serum enzymes than is α-MSH.

III. LIZARD (*ANOLIS*) SKIN BIOASSAY

The American anole (*Anolis carolinensis*), the so-called "American chameleon", is noted for its ability to rapidly change from a very light green color to a very dark brown color.

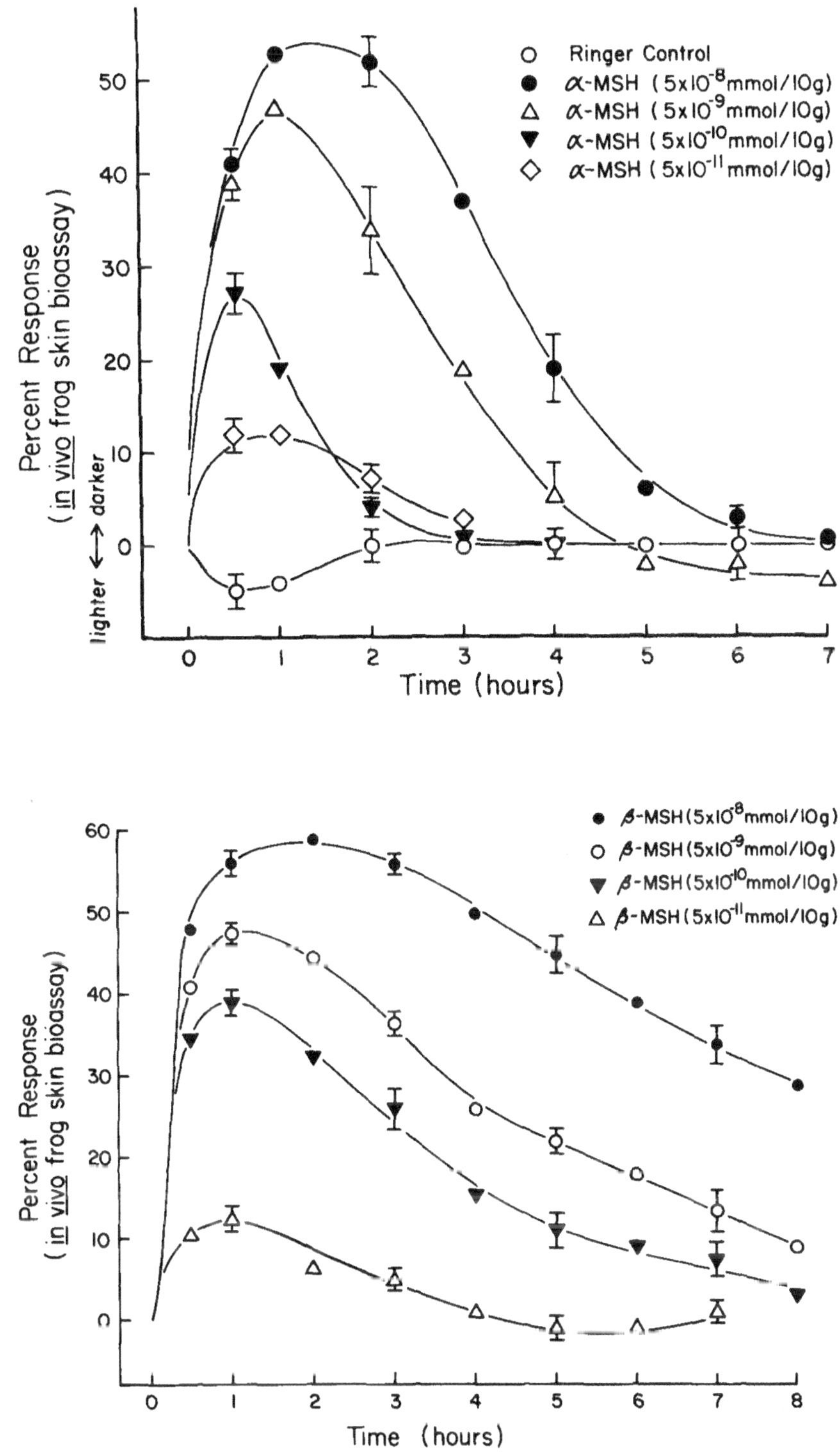

FIGURE 4. In vivo frog (*R. pipiens*) skin bioassay. Darkening response of frogs to (A) α-MSH and (B) β-MSH at several concentrations at the times noted. Each value represents the mean, ± S.E., darkening response of the frogs (n = 10) to the melanotropins at the times noted.

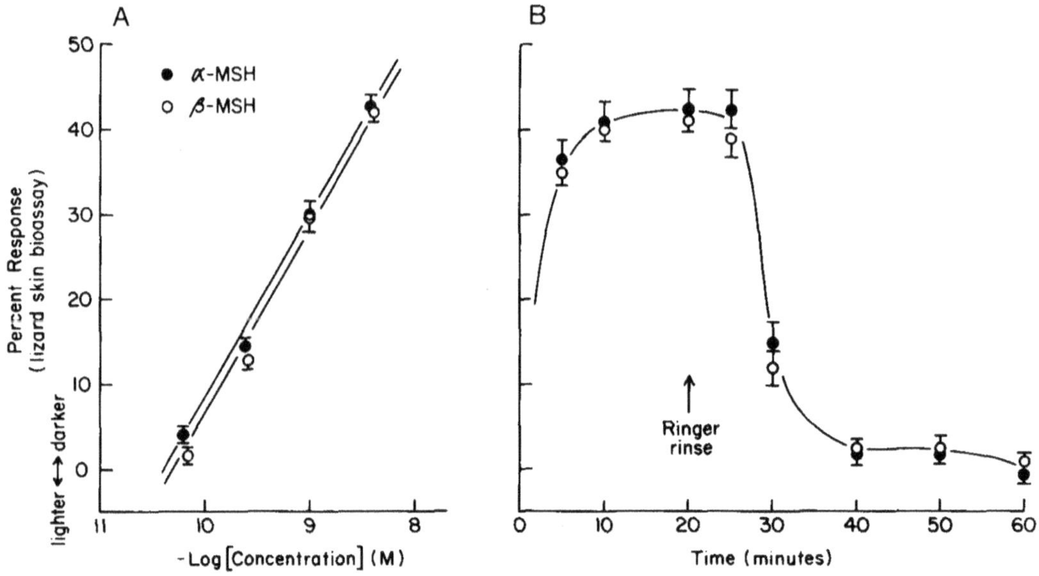

FIGURE 5. In vitro lizard *Anolis*) skin bioassay. (A) Relative dose-responses to α- and β-MSH. (B) Reversal of the darkening of lizard skins after transfer of the skins to Ringer in the absence of the melanotropins. Each value represents the mean, ±S.E., darkening response of the skins (n = 24) to the melanotropins at the concentrations noted.

Small pieces of *Anolis* skin can be floated on Ringer solution wherein they are light green in color; when melanotropins are added to the bathing solution the skins change to an olive green color and then to a light tan color, to a brown color, and finally to a dark brown color. These changes can conveniently be numerically ranked as in the classical MI from one (light green) to five (dark brown).

We have routinely used *Anolis* skins as described for the classical frog skin bioassay.[1] The results of many experiments reveal that α- and β-MSH are equipotent in melanotropic activity in this bioassay (Figure 5 A).[2] This lizard skin assay, although approximately one order of magnitude less sensitive than the frog skin bioassay, is particularly useful since the darkening response to melanotropins and the subsequent lightening response after their removal are very rapid events (Figure 5 B). The skins can be used many times in one day, yielding very similar dose-response curves (Figure 6).[2]

IV. TELEOST (FISH) SKIN BIOASSAY

Chromatophores present within the skin of teleost fishes have been used to evaluate the actions of many melanotropic substances. Erythrophores (red-colored pigment cells) present within scales of the European minnow, *Phoxinus laevis*, were, in fact, used to detect the presence of melanotropic peptides which are apparently elevated during pregnancy. This early "pregnancy" test was, therefore, a bioassay for melanotropic peptides. The responses of melanophores present within isolated scales or fins to hormones or other agents are now generally monitored by photometric methods[11] rather than by the classical MI.

As in amphibians and reptiles, MSH and related peptides stimulate melanosome dispersion within teleost melanophores. Since melanosomes within fish melanophores are dispersed in the so-called "resting" (unstimulated) state, then these intracellular organelles must first be made to aggregate before they can then be stimulated to disperse by MSH-like peptides.

Several fish skin preparations have now been developed which are similar to those used in the frog and lizard skin bioassays as described earlier.[12,13] The fish skin bioassay is unique

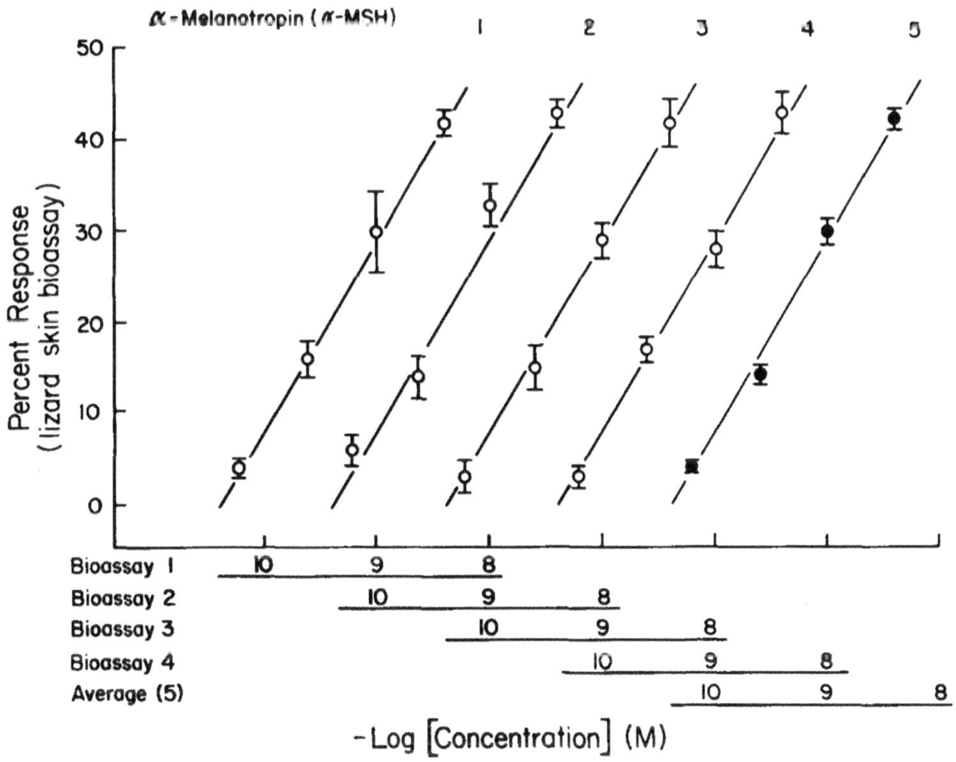

FIGURE 6. In vitro *Anolis* skin bioassay. Lizard skins were alternatively subjected to α- and β-MSH for eight successive bioassays during a 10 hr period of time. The four individual bioassays for α-MSH are shown. Each value represents the mean, ±S.E., darkening repsonse of the skins (n = 6) to MSH at the concentrations indicated (four-fold) dilutions of the biggest concentration employed (4 × 10^{-9} *M*). The mean, ±S.E., reflectance values of the skins (n = 24) in the four bioassays are also shown.

in that it can measure the melanin-aggregating activity of a putative melanin-concentrating hormone (MCH) (see Baker, Volume II, Chapter 11). As in the tetrapod skin bioassays, skins are removed in large pieces and changes in coloration (reflectance) measured by a photometric method. So far, this bioassay has been restricted to only a few species of the teleosts. Other species of smooth-skinned teleosts will certainly be shown to be as useful as the present species now utilized. A tropical eel, *Synbranchus marmoratus*, obtained from the Pantanal ("Big Swamp") in Brazil, has proven to be a good model system for the bioassay of melanotropic peptides. Within a single experiment using several fish, complete dose-response curves can be obtained (Figure 7.)[14] The minimal effective dose of MCH (and MCH_{5-17}) to lighten *Synbranchus* skins is 10^{-12} *M* making this an even more sensitive assay for MCH than even the classical in vitro frog skin bioassay is for MSH. If skins are near-maximally lightened, then the MSH-like activities of peptides can be clearly documented (Figure 7).[14] The receptor specificity of the response to MCH is clearly documented by the lightening response to norepinephrine, that can be blocked by an α-adrenoceptor antagonist which does not compromise the response to MCH.[13]

V. MELANOMA BIOASSAYS

Cancerous melanocytes, melanoma cells, of mice can be propagated in syngeneic mice by serial transplantation of small tissue samples from tumors (melanomas). Tumors can also be broken down into individual cells and the cells grown in culture. Following primary

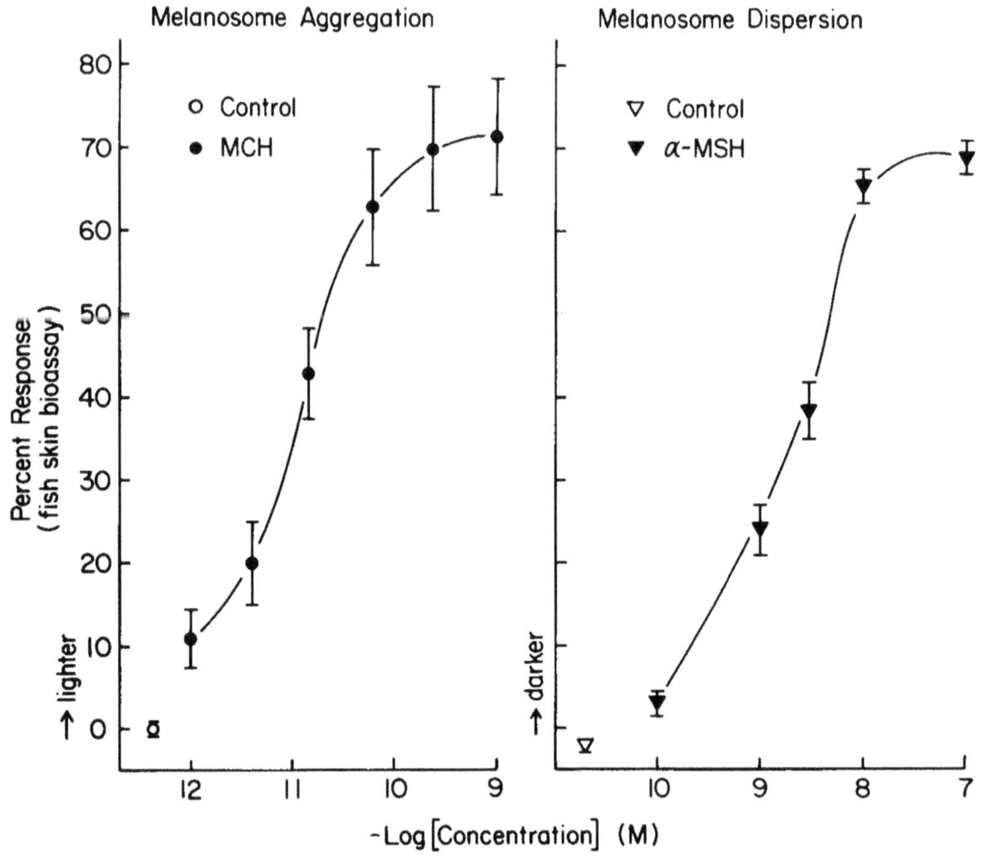

FIGURE 7. In vitro teleost (fish) skin bioassay. Relative potencies of MCH and MSH. Each value represents the mean, ±S.E., lightening or darkening response of the skins (n = 10) to the melanotropic peptides at the concentrations noted. The means, ±S.E., of control groups of skins are also shown.

passage of melanoma cells in culture, the cells can be frozen in liquid nitrogen to provide an almost inexhaustible supply of cells.

The Cloudman S91 melanoma cell line has provided a particularly useful model for the study of melanotropic peptides. The long-term response of S91 melanoma cells to melanotropins can readily be determined by the tyrosinase bioassay. Stimulation of melanoma cell proliferation by melanotropic peptides can also be utilized as an assay for these peptides.[15]

A. Melanoma Tyrosinase Bioassay

Tyrosinase activity is determined by the charcoal absorption method of Pomerantz.[16] This bioassay is based upon the measurement of the amount of ^3H_2O released during the conversion of [^{3}H]tyrosine to L-DOPA, a reaction catalyzed by the enzyme, tyrosinase. The amount of radiolabeled water produced is then monitored by scintillation methods. Although this bioassay is not as sensitive to MSH as either the frog or lizard skin bioassays, the results obtained are quite reproducible. After many passages in culture, however, hormone-stimulated tyrosinase activity may decline. Earlier passage cultures can then be obtained from frozen stocks.

The Cloudman S91 (3960, CCL 531) melanoma cell line can be obtained from the American Type Culture Collection Cell Repository. Cells are grown in Ham's F-10 medium supplemented with 10% heat-inactivated (56°C, 30 min) horse serum and 2% heat-inactivated fetal calf serum. Penicillin and streptomycin (100 units/mℓ, 100 μg/mℓ, respectively) are

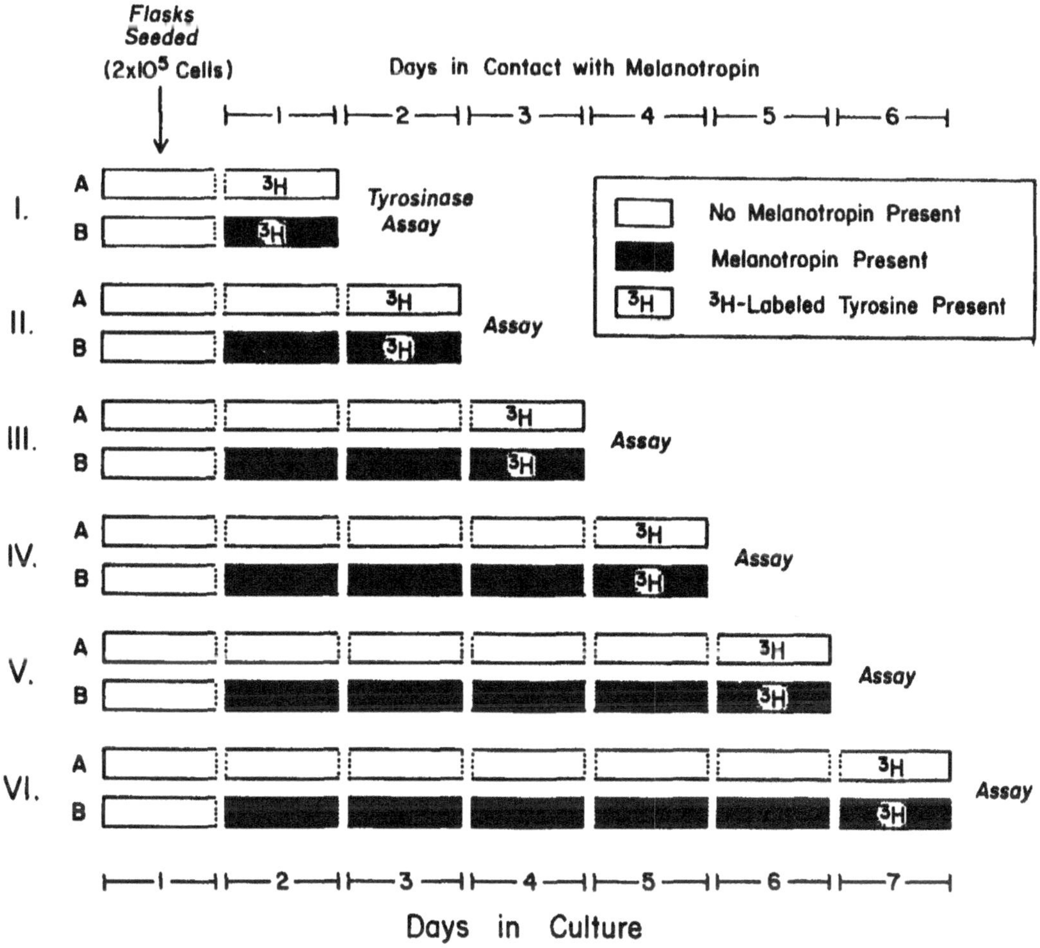

FIGURE 8. Detailed protocol for the determination of melanotropin-stimulated tyrosinase activity over time.

added to the medium. Melanoma cells are incubated at 37°C in a humid atmosphere of 5% CO_2 and 95% air.

For each experiment cells are seeded at a density of 2×10^5 cells per 25-cm^2 flask in 4 mℓ of medium. The number of cells initially seeded will affect the time to reach confluency, at which time the experiment must be terminated. Cells seeded at lower densities than described seem to be retarded relative to their proliferation rate.[17]

In a typical experiment, the cells are seeded into the flasks and allowed to remain 24 hr, during which time they attach to the surface of the flasks. During the subsequent 24 hr, the culture medium is replaced with medium containing 1 μCi/mℓ [^{3}H] tyrosine and α-MSH at one or more concentrations as described. Control flasks also receive the [^{3}H]-labeled tyrosine, but in the absence of the melanotropin. For longer exposures (e.g., 48 and 72 hr), the cells are allowed to attach overnight, and the medium is removed and replaced with medium containing α-MSH each day thereafter. Control flasks receive daily changes of melanotropin-free medium. Then, 24 hr before termination of the experiment, the media in all flasks are replaced with medium containing [^{3}H] tyrosine either alone (control flasks) or in combination with the melanotropin (Figure 8). At the end of the 24-, 48-, and 72-hr exposures, the medium from each flask is collected and assayed for tyrosinase activity. Cells from each flask are also harvested and counted with a hemacytometer. In general, tyrosinase activity is expressed per 10^6 cells, and compared to control tyrosinase activity taken as 100%.

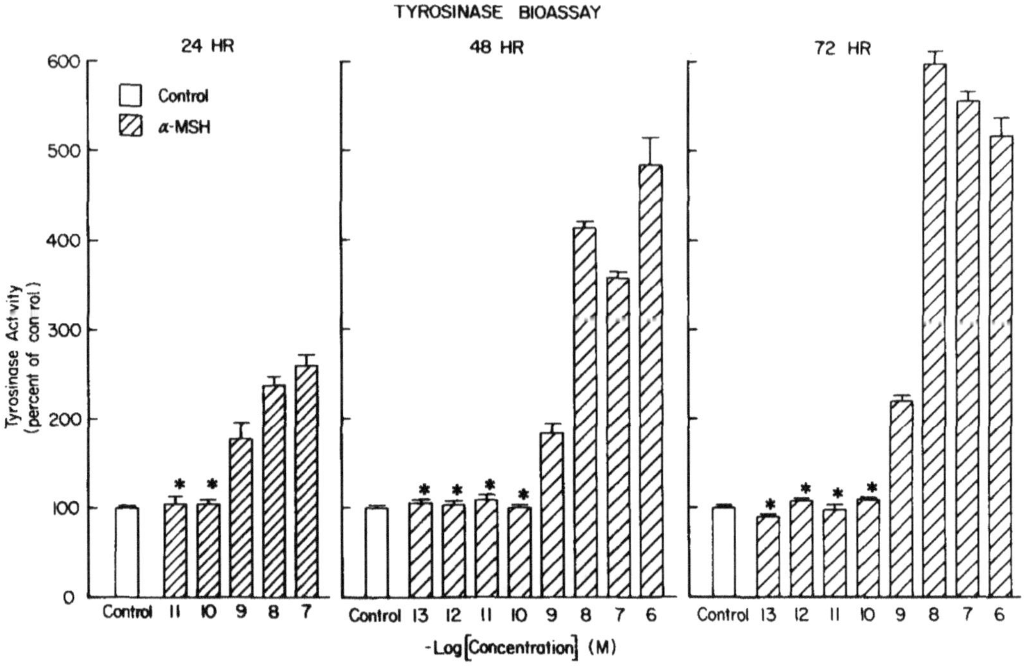

FIGURE 9. Melanoma tyrosinase bioassay. Dose-related stimulation of melanoma tyrosinase activity by MSH at 24, 48, and 72 hr of incubation.

We have consistently determined that the minimal effective dose of α-MSH to significantly stimulate tyrosinase above control levels is between 10^{-10} and 10^{-9} *M* (Figure 9). Certain analogs of α-MSH exhibit significant melanotropic activity at concentrations as low as 10^{-11} to 10^{-12} *M*.[18] Although these analogs are superpotent in other bioassays discussed earlier, they exhibit their greatest potency relative to α-MSH in the tyrosinase bioassay.

Even though (Nle^4,D-Phe^7)-substituted analogs are resistant to inactivation by serum enzymes (see Castrucci et al., Volume I, Chapter 10), the superpotency of these analogs is related to structural and conformational properties of the peptides rather than resistance to enzymes since α-MSH is also unaffected by heat-inactivated sera.[18]

B. Melanoma Adenylate Cyclase Assay

In this assay, the early actions of melanotropins are determined by the use of membrane fragments derived from viable S91 melanoma cells obtained either from tumors or from cultured melanoma cells. In response to melanotropins, membrane-associated adenylate cyclase is activated as determined by the production of ^{32}P-labeled cyclic AMP derived from ^{32}P-labeled ATP substrate (Figure 10). Calcium ions are required for the actions of melanotropins on melanocytes of all vertebrates studied, including S91 melanoma cells. Although Ca^{2+} is required in some undefined process related to melanotropin receptor signal transduction, Ca^{2+} is also inhibitory to the activity of adenylate cyclase. Using membrane preparations from broken cells, it would be expected that both the outer as well as the inner (cytosolic) side of the membrane would be simultaneously exposed to Ca^{2+} present in the incubation medium. This may partly explain why the assay is so insensitive relative to other melanotropin bioassays.[19] Nevertheless, the S91 adenylate cyclase assay gives clear dose-response curves, and the relative potencies of α-MSH analogs are quite similar to those obtained by the frog and lizard skin bioassays.[20] Although basal adenylate cyclase may differ between membrane preparations, it is generally possible to obtain approximately three to

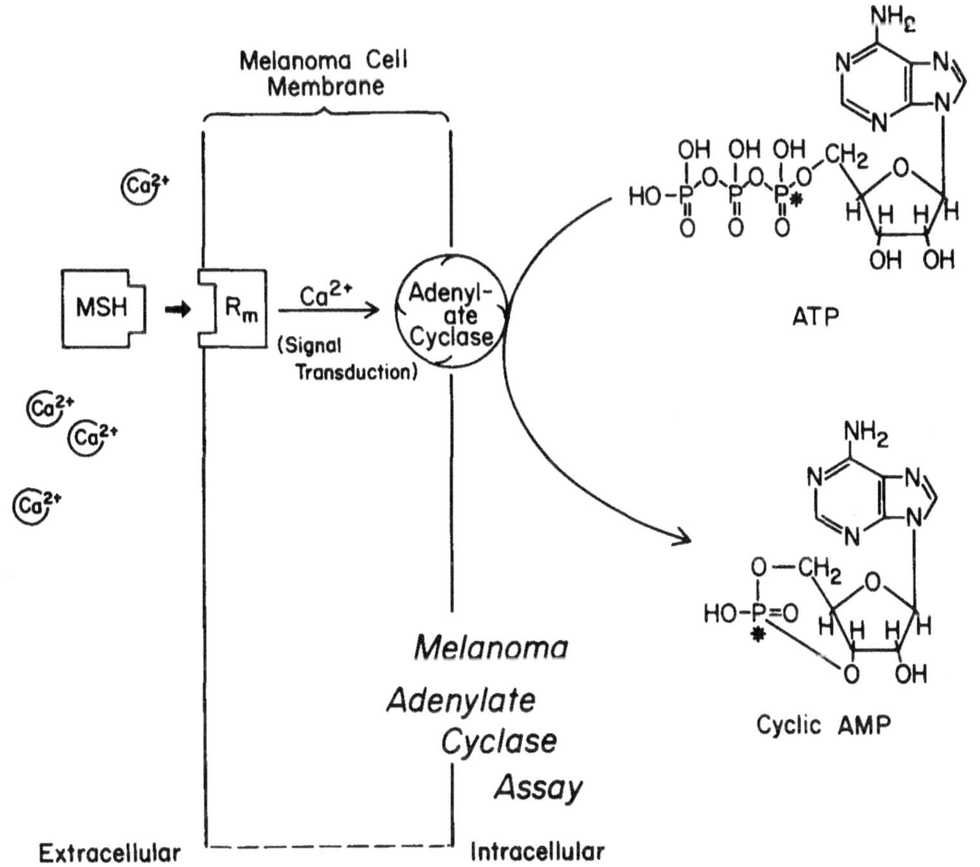

FIGURE 10. Melanoma adenylate cyclase bioassay. The assay measures the amount of labeled cyclic AMP generated in response to a melanotropin (or other agonist) from ^{32}P-labeled ATP substrate.

four significantly separate points with the linear part of the dose-response curve between approximately 10^{-7}- and 10^{-6} *M* concentration of α-MSH.

C. Melanoma Clonogenic Assay

Although a large number of studies have been directed toward the possible use of melanotropins for the destruction of melanoma cells, the evidence is clear that melanotropic peptides stimulate, rather than inhibit, growth of melanoma cells in vitro or in vivo.[15] It has been shown in the S91 melanoma cell clonogenic bioassay that melanotropins stimulate melanoma cell proliferation in a dose-related manner. Therefore, this assay can be utilized to determine the minimal effective dose of melanotropins to stimulate cellular proliferation (see Abdel-Malek, Chapter 3, Volume II).

D. *In Situ* Melanin Assay

This bioassay is based upon the amount of melanin produced by B16 melanoma cells in culture after exposure to melanotropins for 3 days. The cells are plated on 96-well Costar trays, and after incubation with the hormones, the total melanin formed is determined with an automatic Elisa reader, at 405 nm of wavelength. Studies with α-MSH analogs[21] showed that this assay is almost as sensitive to melanotropins as the frog skin bioassay, and approximately two to ten times more sensitive than the tyrosinase bioassay. This assay should also be useful for studies on other melanoma cell types.

VI. PROLONGED (RESIDUAL) ACTIONS OF MELANOTROPINS

The in vitro frog skin bioassay has been criticized because it was reported that melanophores failed to return to the fully contracted state after removal of the melanotropins from the bathing solution. The prolonged dispersion was stated to be due to a "paralysis" of the malanophores from using too high a concentration of melanotropin.[22] A more reasonable explanation for this phenomenon is now available. It is well documented that melanotropins exhibit prolonged (so-called "residual") activity both in vitro and in vivo. We reported that exceptionally prolonged activity could be designed into certain analogs of MSH by replacing L-phenylalanine at position 7 by a D-phenylalanine residue.[23] These peptides exhibit superpotency in vitro and extraordinary prolongation of melanotropic activity both in vitro and in vivo (see Sawyer et al., Chapter 5, this Volume).

It is reasonable to assume that synthetic peptides and even peptides from natural sources will be contaminated to some degree with D-amino acid residues. Therefore, when melanotropins are used at very high, unphysiological concentrations, even a very small amount ($\leq$0.1%, or less) of the contaiminating D-phenylalanine[7] diastereomer will cause prolonged action of melanocytes. Since D-Phe[7]-containing melanotropins are resistant to inactivation by serum enzymes, these peptides will also exhibit superpotency in vivo (see Castrucci et al.,[24] Chapter 10, Volume I).

VII. PERSPECTIVES

The structural requirements for α-MSH activity have been very well defined. Deletions and substitutions of amino acid residues within the primary structure of α-MSH have determined the essential elements required for potency, superpotency, and prolonged (residual) biological activity (see Chapter 5, this Volume). Over 300 analogs have been synthesized and their biological activities determined. The ability to study the structural and conformational requirements for α-MSH activity is due largely to the fact that several simple, as well as more sophisticated bioassays are available for such determinations. These assays provide rather definitive formation on the rapid, as well as the long-term (genomic) actions of melanotropins.

ACKNOWLEDGMENTS

These studies were supported in part by grants from Public Health Service (MH-27257, MH-30626, and AM-17420), from the National Science Foundation (PCM-882200, PCM-811220, PCM-770731, and PCM-810078), U.S.A., and from the Fundacao de Amparo a Pesquisa do Estado de São Paulo (84/1967, 85/0718-7), Brasil.

REFERENCES

1. **Shizume, K., Lerner, A. B., and Fitzpatrick, T. B.,** *In vitro* bioassay for melanocyte stimulating hormone, *Endocrinology,* 54, 553, 1954.
2. **Castrucci, A. M. de L., Hadley, M. E., and Hruby, V. J.,** Melanotropin bioassays:*in vitro* and *in vivo* comparisons, *Gen. Comp. Endocrinol.,* 55, 104, 1984.
3. **Huntington, T. and Hadley, M. E.,** Evidence against mass action direct feedback control of melanophore-stimulating hormone (MSH) release, *Endocrinology,* 96, 472, 1974.
4. **Tilders, F. J. H., Delft, A. M. L. Van, and Smelik, P. G.,** Reintroduction and evaluation of an accurate, high capacity bioassay for melanocyte-stimulating hormone using the skin of *Anolis carolinensis in vitro, J. Endocrinol.,* 66, 165, 1975.

5. **Dickhoff, W. W.**, A rapid, high-efficiency bioassay of melanocyte-stimulating hormone, *Gen. Comp. Endocrinol.*, 33, 304, 1977.
6. **Bjorklund, A., Meurling, P., Nilsson, G., and Nobin, A.**, Standardization and evaluation of a sensitive and convenient assay for melanocyte-stimulating hormone using *Anolis* skin *in vitro*, *J. Endocrinol.*, 53, 161, 1972.
7. **Parker, G. H.**, *Animal Colour Changes and Their Neurohumours*, Cambridge University Press, London, 1948.
8. **Hogben, L. T. and Slome, D.**, The pigmentary effector system. VI. The dual character of endocrine co-ordination in amphibian colour change, *Proc. R. Soc. London Ser. B*, 108, 10, 1931.
9. **Sawyer, T. K., Hruby, V. J., Wilkes, B. C., Draelos, M. T., Hadley, M. E., and Bergsneider, M. E.**, Comparative biological activity of highly potent active-site analogues of α-melanotropin, *J. Med. Chem.*, 25, 1022, 1982.
10. **Kastin, A. J., Schally, A. V., Yajima, H., and Kubo, K.**, Melanocyte stimulating hormone activity of synthetic MSH and ACTH peptides, *in vivo* and *in vitro*, *Nature (London)*, 207, 978, 1965.
11. **Oshima, N., Kasukawa, H., Fujii, R., Wilkes, B. C., Hruby, V. J., Castrucci, A. M. de L., and Hadley, M. E.**, Melanin concentrating hormone (MCH effects on telost *(Chrysiptera cyanea)* melanophores, *J. Exp. Zool.*, 235, 175, 1985.
12. **Wilkes, B. C., Hruby, V. J., Sherbrooke, W. C., Castrucci, A. M. de L., and Hadley, M. E.**, Synthesis and biological actions of melanin concentrating hormone, *Biochem. Biophys. Res. Commun.*, 122, 615, 1984.
13. **Wilkes, B. C., Hruby, V. J., Castrucci, A. M. de L., Sherbrooke, W. C., and Hadley, M. E.**, Synthesis of a cyclic melanotropic peptide exhibiting both melanin-concentrating and-dispersing activities, *Science*, 224, 1111, 1984.
14. **Castrucci, A. M. de L., Hadley, M. E., and Hruby, V. J.**, A teleost skin bioassay for melanotropic peptides, *Gen. Comp. Endocrinol.*, 66, 374, 1987.
15. **Abdel Malek, Z. A., Hadley, M. E., Bregman, M. D., Meyskens, F. L., Jr., and Hruby, V. J.**, Actions of melanotropins of mouse melanoma cell growth *in vitro*, *J. Natl. Cancer Inst.*, 76, 857, 1985.
16. **Pomerantz, S. H.**, The tyrosine hydroxylase activity of mammalian tyrosinase, *J. Biol. Chem.*, 241, 161, 1966.
17. **Abdel Malek, Z. A., Kreutzfeld, K. L., Hruby, V. J., Bregman, M. D., Meyskens, F. L., Jr., and Hadley, M. E.**, Melanotropin stimulation of S91 melanoma cell tyrosinase is density dependent, *In Vitro*, 22, 75, 1986.
18. **Marwan, M. M., Abdel Malek, Z. A., Kreutzfeld, K. L., Hadley, M. E., Wilkes, B. C., Hruby, V. J., and Castrucci, A. M. de L.**, Stimulating of S91 melanoma tyrosinase activity by superpotent α-melanotropins, *Mol. Cell. Endocrinol.*, 44, 171, 1985.
19. **Kreiner, P. W., Gold, C. J., Keirns, J., Brock, W. H., and Bitensky, M. W.**, MSH-sensitive adenylyl cyclase in the Cloudman melanoma, *Yale J. Biol. Med.*, 46, 583, 1973.
20. **Bregman, M. D., Sawyer, T. K., Hadley, M. E., and Hruby, V. J.**, Adenosine and divalent cation effects of S91 melanoma adenylate cyclase, *Arch. Biochem. Biophys.*, 200, 1, 1980.
21. **Siegrist, W., Stutz, S., Girard, J., and Eberle, A. N.**, A new bioassay for MSH: *in situ* melanin assay for melanoma cells in culture, in Abstr. from Satellite Conf. on Melanotropic Peptides, 1986.
22. **Hudson, B. and Bentley, G. A.**, The biological assay of melanocyte stimulating hormone, *Austral. J. Exp. Biol.*, 35, 45, 1957.
23. **Sawyer, T. K., Sanfillippo, P. J., Hruby, V. J., Engel, M. H., Heward, C. B., Burnett, J. B., and Hadley, M. E.**, 4-Norleucine, 7-D-phenylalanine-α-melanocyte-stimulating hormone: a highly potent α-melanotropin with ultraprolonged biological activity, *Proc. Natl. Acad. Sci. U.S.A.*, 77, 5754, 1980.
24. **Castrucci, A. M. de L., Hadley, M. E., Yorulmazoglu, E. I., Wilkes, B. C., Sawyer, T. K., and Hruby, V. J.**, Synthesis and studies of superpotent melanotropins resistant to enzyme degradation, in *Biological, Molecular and Clinical Aspects of Pigmentation*, Bagnara, J., Klaus, S. N., Paul, E., and Schartl, M., Eds., University of Tokyo Press, 1985, 145.

Chapter 2

MELANOTROPIN MECHANISMS OF ACTION: MELANOSOME MOVEMENTS

Mac E. Hadley and Ana Maria de L. Castrucci

TABLE OF CONTENTS

I. Introduction 16
A. Melanocytes, Melanogenesis, Melanosomes 16

II. Melanotropic Peptides and Melanosome Dispersion 16
A. Role of Cyclic Adenosine Monophosphate (cAMP) 16
B. Role of Protein Kinases 17
C. Separate Receptors for Melanotropin Action 17
D. Sulfhydryl Requirement for Melanotropin Action 17
E. Calcium Requirement for Melanotropin Action 18
F. Melanofilaments and Melanotropin Action 19

III. Melanotropic Peptides and Melanosome Aggregation 20

IV. Melanotropic Peptides: Structure-Activity Studies 22
A. Melanocyte Stimulating Hormone (MSH) 22
B. Melanin Concentrating Hormone (MCH) 22

V. Model of Melanotropin Control of Melanosome Movements 23

VI. Summary and Perspectives 24

Acknowledgments 24

References 24

I. INTRODUCTION

α-Melanotropin (α-melanocyte-stimulating hormone or α-MSH, herein referred to as MSH) synthesized and released by the vertebrate pituitary gland, acts on pigment cells within the skin.[1] In response, these effector cells, melanocytes (melanophores), synthesize a pigment, melanin, on a subcellular organelle, the melanosome (melanin granule). Melanin pigmentation of the skin serves a variety of functions depending on the particular adaptive needs of the species (see Weatherhead, Volume II, Chapter 1). One important function of integumental melanocytes in poikilothermic vertebrates is to provide a means for rapid color changes of the skin.[2] Color change is effected, in part, by melanosome movements within melanocytes. The mechanisms by which melanotropic peptides regulate melanosome movements within melanocytes are the topic of this chapter.

A. Melanocytes, Melanogenesis, Melanosomes

Melanotropic peptides play essential roles in the control of melanocytes and melanin pigmentation of the skin in many animals. However, the particular function that melanocytes subserve in the integument is dependent upon whether the neural crest stem (propigment) cells migrate to the dermis or to the epidermis of the skin. Within both the epidermis and the dermis melanogenesis leads to melanosome formation within melanocytes. Melanogenesis and melanosome formation continue as important functions of epidermal melanocytes.[3] Mature melanosomes are continuously released into adjacent keratinocytes of the epidermis and are replaced within the melanocytes by newly synthesized melanosomes. In many animals, melanosome deposition within the epidermis provides the structural element for certain epidermal specializations. In homeotherms, melanosomes deposited within cellular elements of hair and feathers provide the basis for coloration in these epidermal structures.

In dermal melanocytes, in contrast, melanogenesis ceases after a certain number of melanosomes are produced, and these melanosomes are retained by the melanocytes. Dermal melanocytes, like muscle cells, are contractile-like cells within which melanosomes can be translocated rapidly throughout the cells in response to melanotropins or certain other hormones. The mobilization of these pigment organelles within melanocytes provides the cellular basis for rapid color changes that characterize many poikilothermic vertebrates.[2]

II. MELANOTROPIC PEPTIDES AND MELANOSOME DISPERSION

MSH released by the pars intermedia of the pituitary gland is responsible for darkening of the skin of most species of vertebrates that have been studied (see Bagnara, Volume II, Chapter 2). MSH also darkens skins in vitro and, as in vivo, this darkening results from melanosome dispersion within melanocytes. MSH also stimulates melanosome dispersion within isolated melanocytes maintained in tissue culture, thus demonstrating that the actions of the hormone are directly on the pigment cells.[4]

A. Role of Cyclic Adenosine Monophosphate (cAMP)

As for many other cell types, MSH mediates its action through stimulation of the production of cAMP, a so-called "second messenger" of hormone action.[5] cAMP levels are elevated in frog skins darkened by MSH, but not in those areas of the skin lacking melanocytes. Levels of the cyclic nucleotide become lowered if frog skins darkened by MSH are then lightened by the addition of norepinephrine or melatonin, chemical messengers which stimulate melanosome aggregation.[6] Although other chromatophores in addition to melanocytes participate in the darkening response,[2,7,8] it is clear that elevated levels of cAMP are correlated with the response and, therefore, with melanosome dispersion.

Frog skins darken in response to theophylline and other methylxanthines (e.g., caffeine) which are known inhibitors of phosphodiesterase, an enzyme which degrades cAMP within cells.[9] More directly, dibutyryl cAMP (DcAMP), an analog of cAMP which more readily enters cells, also darkens frog skin.[10] Theophylline and DbcAMP also mediate their actions directly on melanocytes maintained in tissue culture. Similar results have been obtained with integumental melanocytes of other vertebrate species, reptiles and fish.[11,12] Taken together, these observations reveal that cAMP regulates melanosome dispersion and is the intracellular mediator of the actions of "first messengers", MSH.

Abnormal melanocytes, melanoma cells, grown in tissue culture, also respond to MSH by an increase in cAMP levels (see Pawelek,[13] Chapter 4, this Volume). Evidence that the immediate action of MSH involves the synthesis of cAMP is derived from studies using membrane fragments from melanoma cells.[14,15] MSH added to melanoma membranes stimulates cAMP formation as shown by the conversion of radiolabeled ATP to radiolabeled cAMP (see Hadley and Castrucci, Chapter 1, this Volume). This observation indicates that the initial actions of MSH on melanoma cells are mediated at the level of the plasma membrane. It might be expected, therefore, that the initial actions of MSH on normal melanocytes also involve the production of cAMP.

B. Role of Protein Kinases

The intracellular actions of cAMP are probably mediated through activation of a cAMP-dependent protein kinase. This kinase may then initiate a cascade of events which involve a number of substrate phosphorylations. Protein kinase activation by MSH in melanoma cells has been demonstrated (see Pawelek, Chapter 4, this Volume). MSH has also been shown to induce specific changes in the degree of phosphorylation of a 53-kDa protein concomitant with melanosome dispersion in isolated frog melanocytes (see de Craan and Eberle,[16] Chapter 3, this Volume). It might be assumed, therefore, that substrate phosphorylation by cAMP is a requisite for melanosome dispersion.[16-18]

C. Separate Receptors for Melanotropin Action

Agonists other than melanotropins also stimulate melanosome dispersion within melanocytes. Catecholamines acting through α-adrenoceptors, stimulate melanosome dispersion, a response that can be blocked by α-adrenoceptor antagonists, such as propranolol.[19,20] The actions of propranolol are specific since melanosome dispersion in response to MSH is not similarly antagonized, Also, sulfhydryl inhibitors block MSH-induced dispersion of melanosomes, but do not inhibit catecholamine-induced dispersion of these organelles.[21] These observations reveal that the individual actions of MSH or a catecholamine can be preferentially blocked without compromising the action of the other agonist. Therefore, cAMP stimulation by MSH is mediated through receptors separate from those through which other agonists similarly activate cyclic nucleotide production (Figure 1).

D. Sulfhydryl Requirement for Melanotropin Action

Mercurials are regarded as being reasonably specific for sulfhydryl groups, but nonspecific as to which tissue sulfhydryl group is attacked. It has been shown, however, that mercurials and some other sulfhydryl inhibitors cause a highly selective inhibition of MSH action of melanocytes.[21] These agents inhibit melanosome dispersion caused by MSH, but they do not interfere with other hormonal- (catecholamine) or pharmacological- (theophylline, DbcAMP) stimulated melanosome dispersion. Since sulfhydryl inhibitors antagonize MSH, but not catecholamine-, methylxanthine-, or DbcAMP-induced darkening, the site of the inhibitory action of these agents may be at the membrane receptor and/or at one of the sequential molecular mechanisms between the melanotropin receptor and adenylate cyclase activation.

These data reveal two separate physiological mechanisms for initiating melanosome dis-

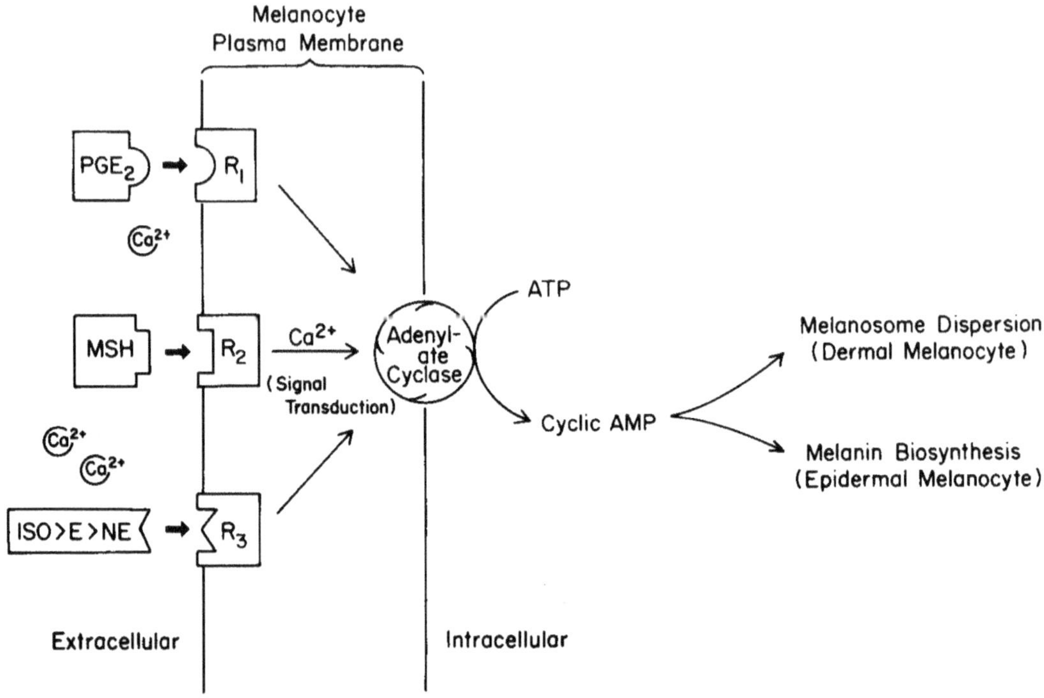

FIGURE 1. Receptor specificity for melanosome-dispersing agonists. Activation of adenylate cyclase is, however, a common step after the activation of MSH, catecholamine, and prostaglandin receptors. ISO, isoproterenol; E, epinephrine; NE, norepinephrine.

persion: (1) an MSH mechanism for which intact sulfhydryl groups are required and which is blocked by sulfhydryl inhibitors and (2) a catecholamine mechanism mediated through β-adrenoceptors which lacks such a requirement. These results also reveal that the sulfhydryl requirement for MSH action is receptor specific and is related to the the initial mechanism of receptor signal transduction and not to the actions of cAMP or to melanosome movements per se. Since melanosome dispersion induced by catecholamines in the presence of sulfhydryl inhibitors can be reversed by catecholamines (following α-adrenoceptor blockade), then the actions of sulfhydryl inhibitors also do not affect the mechanisms related to melanosome aggregation.

E. Calcium Requirement for Melanotropin Action

There is a calcium requirement for MSH action on melanocytes, but not for melanosome dispersion per se since melanosome dispersion proceeds in a Ca^{2+}-free medium in response to methylxanthines or DbcAMP. Calcium is, in fact, the only cation specifically required for MSH action since sodium and potassium ions can be replaced by certain other monovalent cations (e.g., lithium, cesium, and rubidium).[22-24]

The Ca^{2+} requirement for MSH action is receptor specific since prostaglandins (e.g., PGE_2) and catecholamines (acting through β-adrenoceptors) stimulate melanosome dispersion in the absence of the cation (Figure 1).[23] Since Ca^{2+} is not required for the intracellular mechanisms controlling melanosome dispersion, then it is likely that Ca^{2+} is required for transduction of the signal between the MSH receptor and activation of adenylate cyclase. Calcium could be required for MSH receptor interaction or for some process between receptor and adenylate cyclase activation. We have provided evidence elsewhere that the latter suggestion is more likely to be correct.[25]

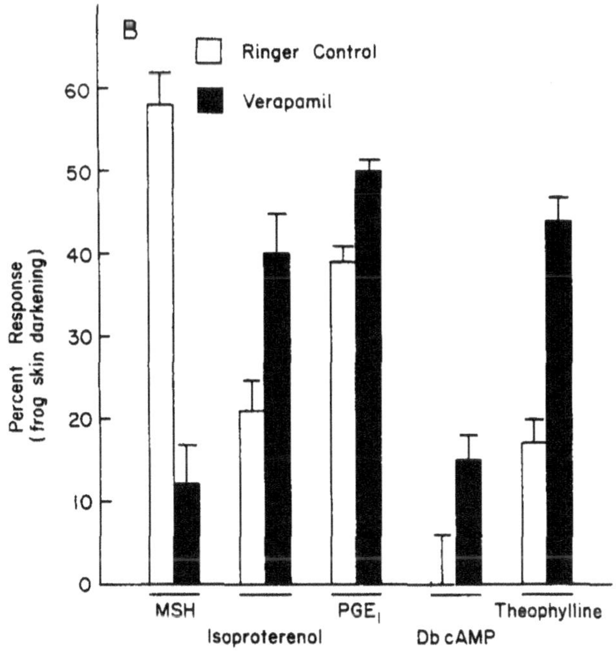

FIGURE 2. Specific blockade of MSH darkening actions by Verapamil, whereas the darkening activity of other agonists is not compromised in the presence of the Ca^{2+} channel antagonist, as demonstrated by the in vitro frog (*Rana pipiens*) skin bioassay. PGE_1, prostaglandin E_1; DbcAMP, dibutyryl cyclic AMP.

Verapamil (Isoptin®), a Ca^{2+} channel antagonist, blocks MSH-induced melanosome dispersion. The actions of prostaglandins and catecholamines (acting through β-adrenoceptors) are not blocked by the drug. In the presence of verapamil, the actions of the other melanosome-dispersing agonists are actually enhanced compared to their actions in the absence of the drug.[26,27] These results again emphasize the receptor-specific requirement of Ca^{2+} for MSH action (Figure 2).[23]

Lithium, choline, rubidium, and cesium, in the lizard and the toad, will replace the sodium and potassium of the Ringer medium and permit melanosome dispersion, but only if Ca^{2+} is present. These monovalent cations are irreversibly inhibitory to MSH action on melanophores of *Rana berlandieri* and *R. pipiens*. These results demonstrate that the Ca^{2+} is the only cation specifically required for MSH action on melanophores.[24]

F. Melanofilaments and Melanotropin Action

Melanocytes are typically characterized by an array of filamentous organelles. Considering the ability of melanosomes to be rapidly translocated from the cell center out into the dendritic processes of the cells and back again, it might be expected that melanofilaments of one or more types participate in the actions of MSH related to melanosome dispersion. Colchicine induces dispersion of amphibian[28] and fish melanosomes, and blocks the actions of agonists which normally reverse the actions of MSH.[29] Cytochalasin B, on the other hand, blocks the actions of MSH on melanocytes.[28,29] Colchicine and cytochalasin B inhibit or otherwise antagonize the integrity or function of microtubules and microfilaments, respectively. Microinjections of antiactin antibody completely block both dispersion and aggregation in fish melanophores.[30] It has, therefore, been concluded that melanosome dispersion, at least in amphibian and fish melanophores, in response to MSH, involves the active participation of

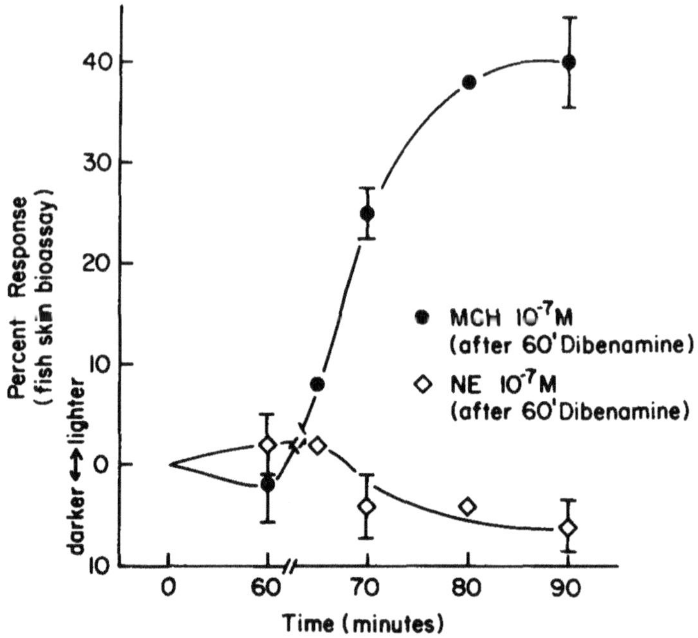

FIGURE 3. Receptor specificity for MCH lightening activity in the in vitro fish (*Gobeisox pinniger*) skin bioassay. Dibernamine blocks norepinephrine melanosome-aggregating activity, whereas MCH actions are not compromised.

microfilaments, whereas melanosome aggregation may require the participation of microtubules.[28,29] Possibly one or more types of melanofilament proteins, such as actin and dynein-tubulin system, may serve as the substrates for a protein kinase that may be activated by MSH.[16,17] A microtrabecular lattice has been proposed to explain why melanosomes travel in a coordinate fashion.[31] This three-dimensional network has been visualized with high resolution microscopy in fish chromatophores.[32] During aggregation, the trabecular lattices become shorter and thicker, and during dispersion they are expanded along microtubules.[33]

Evidence points out that pigment dispersion and aggregation rely on different mechanisms, but certainly depend on the integrity of cytoskeleton elements. Whatever the mechanism is, the motile force that generate pigment translocation, is still poorly understood.

III. MELANOTROPIC PEPTIDES AND MELANOSOME AGGREGATION

A peptide exhibiting melanosome-aggregating activity, a so-called "melanin-concentrating hormone" (MCH), was isolated from the fish pituitary gland (see Baker, Volume II, Chapter 11). The primary structure was determined and the peptide was synthesized by several laboratories.[34,35] This heptadecapeptide stimulates melanosome aggregation within teleost melanocytes at nanogram or lower concentrations.[34] Unlike in tetrapods (e.g., amphibians and reptiles), melanosomes are dispersed within teleost melanocytes in the "resting" (unstimulated) state. In all teleost species so far studied, MCH will lighten the skin by inducing melanosome aggregation. Skins lightened by MCH will then darken in response to MSH.[35]

The actions of MCH are mediated through receptors from those through which other agonists such as catecholamines or melatonin also induce melanosome aggregation. For example, the actions of norepinephrine are blocked by the α-adrenoceptor antagonist, Dibenamine, whereas the actions of MCH are unaffected (Figure 3).[35]

It was noted earlier that the actions of MSH in stimulating melanosome dispersion within

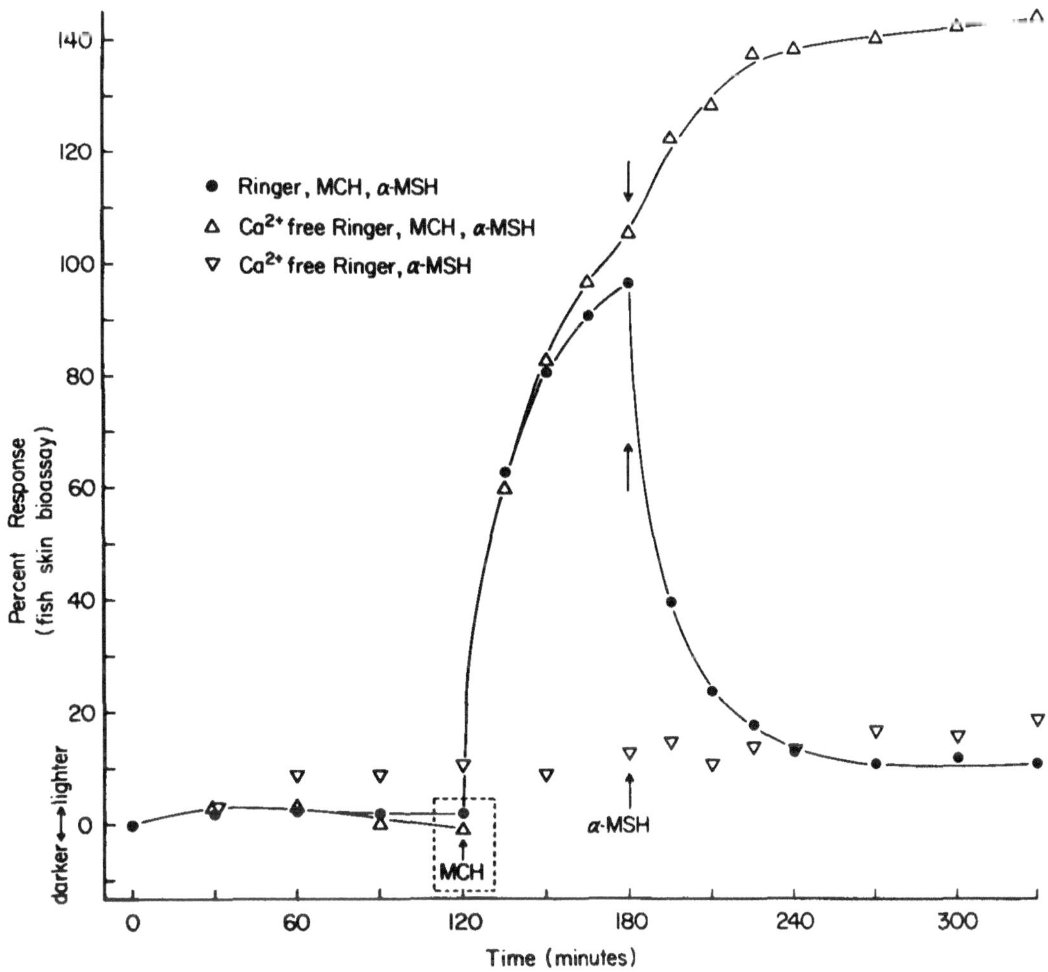

FIGURE 4. Lack of calcium requirement for MCH lightening actions, whereas MSH darkening activity is totally abolished in the absence of the cation, as demonstrated by the in vitro is totally abolished in the absence of the cation, as demonstrated by the in vitro fish (*Synbranchus marmoratus*) skin bioassay.

frog and lizard melanocytes are Ca^{2+} dependent, but that organelle dispersion per se is without such a requirement. Calcium is not required for the actions of MCH, a result which agrees with the observation that other melanosome-aggregating agonists can mediate their actions in the absence of extracellular Ca^{2+}. The lack of a Ca^{2+} requirement for MCH action is quite specific as shown by the inhibition of MSH-induced darkening (melanosome dispersion) of fish skins in a Ca^{2+}-free medium (Figure 4).[36]

These observations reveal that two peptide hormones, MSH and MCH, exhibit opposing actions on melanosome movements within teleost melanocytes (Figure 5). Since it would not be expected that in nature two hormones could exsert contrasting actions through the same receptor, then it would seem reasonable that the receptors through which the two melanotropins exert their actions are separate. This suggestion is further strengthened by the observations that the actions of MSH, but not those of MCH, are Ca^{2+} dependent (Figure 4). In addition, if the actions of both peptide hormones were mediated through the same receptor, then it would be expected that MCH could also reverse the action of MSH on frog and lizard skin melanocytes as it does on teleost chromatophores. MCH does not reverse the actions of MSH on tetrapod melanocytes; the peptide only causes melanosome dispersion.[34] These results suggest that tetrapod melanocytes lack MCH receptors.

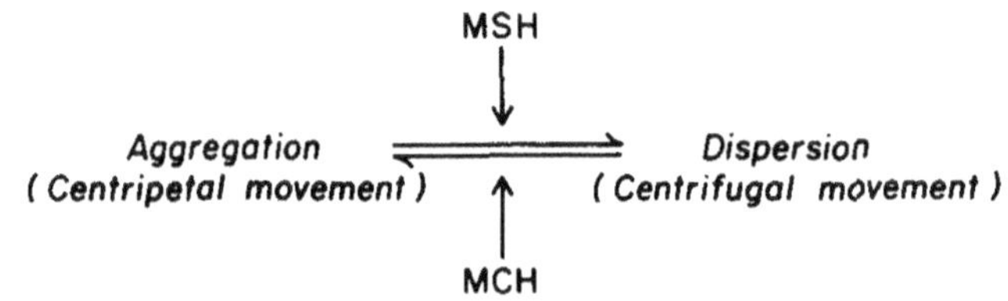

FIGURE 5. MSH and MCH exhibit opposing actions on melanosome movements (dispersion and aggregation, respectively) within teleost melanocytes.

IV. MELANOTROPIC PEPTIDES: STRUCTURE-ACTIVITY STUDIES

Bioassays using frog, lizard, and more recently, fish skins, as well as melanoma cells grown in culture, have provided a substantial body of data relative to the structural requirements of melanotropic peptides necessary to stimulate melanosome movements within melanocytes.

A. Melanocyte-Stimulating Hormone (MSH)

α-MSH is a tridecapeptide, Ac-Ser-Tyr-Ser-Met-Glu-His-Phe-Arg-Trp-Gly-Lys-Pro-Val-NH_2, which shares a central heptadecapeptide sequence, Met-Glu-His-Phe-Arg-Trp-Gly-Lys, with several other melanotropins, including α-MSH and adrenocorticotropin (ACTH). Only α-MSH, however, is considered to be the physiologically relevant systemic hormone regulating integumental melanocytes. Classical structure-activity studies derived particularly from frog and lizard skins, reveal that the minimal sequence to stimulate melanosome dispersion is the tetrapeptide sequence, His-Phe-Arg-Trp.[37] Although this sequence is quite inactive, as might be expected, it apparently represents the minimal natural sequence required for melanotropin receptor binding, receptor signal transduction, and cAMP production leading to melanosome dispersion. In the frog, MSH_{4-12} represents the shortest fragment analog that will give a near equipotent response to the native hormone, MSH. In the lizard, MSH_{4-11} is only slightly less active than MSH. These observations reveal that melanotropin receptors differ between species.

These results also demonstrate that the N-terminal tripeptide sequence is relatively unimportant for receptor activation and cAMP production. The C-terminal valine moiety is also nonessential for the melanotropic activity of α-MSH. The receptors between frogs and lizards differ considerably relative to a need for position 12 (proline) of MSH for activity.[37]

B. Melanin Concentrating Hormone (MCH)

It was noted earlier that MCH is a heptadecapeptide, Asp-Thr-Met-Arg-Cys-Met-Val-Gly-Arg-Val-Tyr-Arg-Pro-Cys-Trp-Glu-Val, which is probably synthesized by the hypothalamus and stored and secreted by the teleost pituitary gland. This peptide may be responsible for controlling the lightening response of some fishes. This melanotropic peptide is very potent (MED, 10^{-12} *M*) in stimulating melanosome aggregation within teleost melanocytes.[38]

Several fragment analogs of MCH have been synthesized and characterized for melanotropic activity: MCH_{1-14}, MSH_{5-17}, and MCH_{5-14}.[38] Structure-activity studies have determined that the 5-17 sequence was as active as the native peptide, suggesting that the N-terminal tetrapeptide sequence may be of no importance for receptor activation. The 1-14 sequence was approximately 30 times less active than the native 1-17 sequence, and the central cyclic 5-14 sequence proved to be approximately 300 times less active. One or more residues of the 15-17 sequence are apparently essential for equipotency at the MCH receptor. At higher,

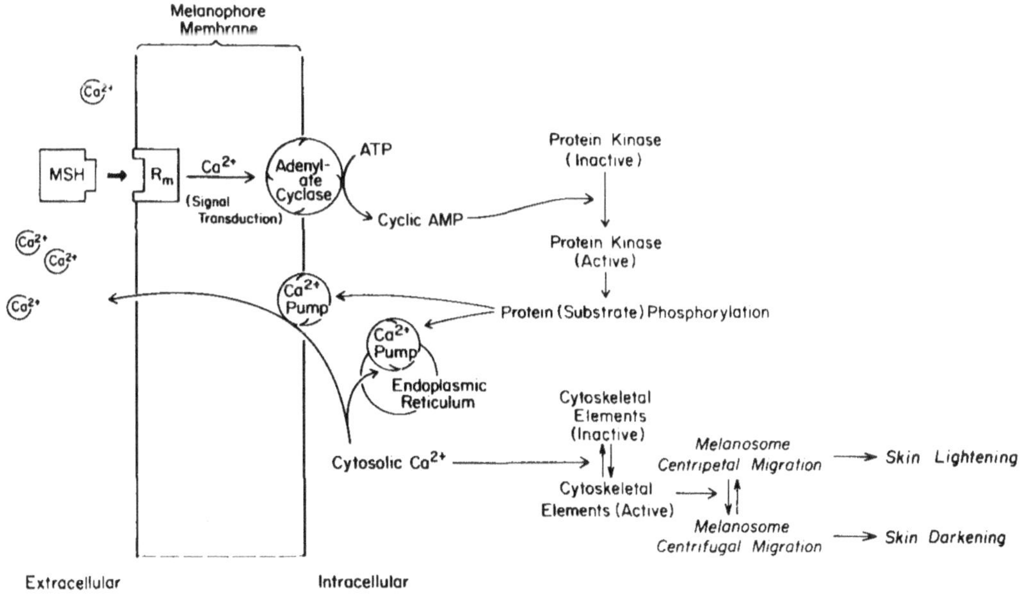

FIGURE 6. Hypothetical model for MSH-induced melanosome dispersion within melanocytes.

unphysiological concentrations, MCH exhibited a self-antagonism, an autoinhibition.[39] It was hypothesized that the peptide possessed some intrinsic MSH-like activity and that MCH and MSH might be evolutionarily related (see Sherbrooke et al., Volume II, Chapter 12).

V. MODEL OF MELANOTROPIN CONTROL OF MELANOSOME MOVEMENTS

Based upon the information discussed above, the following model for the control of melanosome movements within melanocytes by melanotropic peptides is proposed:

> There is a bifunctional role for Ca^{2+} in the control of melanosome dispersion within melanocytes in response to MSH. Calcium is clearly necessary for the activation of adenylate cyclase at the level of the plasma membrane. The cAMP produced within the cytosol may then activate a protein kinase which is responsible for phosphorylating one or more substrate proteins. One of these proteins may be a component of a Ca^{2+} pump localized to the plasma membrane and/or some cytoplasmic organelle (e.g., melanoplasmic reticulum). Activation of this pump might then lower the level of cytosolic Ca^{2+}, either by extruding the cation out of the cell or by causing Ca^{2+} to be sequestered to some cellular component. Lowered levels of cytosolic Ca^{2+} might then trigger the activation of one or more types of melanofilaments which may provide the motile force for melanosome dispersion (Figure 6).[40]

Presumably, melanosome aggregation would then take place following an increase in cytosolic levels of Ca^{2+} following termination of MSH action. Ca^{2+} is not required for the initial actions of agonists that antagonize MSH (e.g., norepinephrine, melatonin, and MCH). Whether the actions of MCH (and other melanosome-aggregating agents) involve an increase in an intracellular second messenger and/or a decrease in the intracellular levels of cAMP is unknown.

VI. SUMMARY AND PERSPECTIVES

From studies on one or more species of most classes of poikilothermic vertebrates (cyclostomes, elasmobranchs, teleosts, amphibians, reptiles, and mammals), it can be concluded that MSH stimulates melanosome dispersion. In contrast, the melanosome-aggregating actions of MCH may be restricted to teleosts. MCH is without such effect on tetrapods (frog and lizard) and the few primitive fish that have been studied (see Sherbrooke et al., Volume II, Chapter 12). In those species of animals that lack MCH receptors, the normal lightening response may simply result from a cessation of MSH secretion. Except for teleost fishes, it appears that color changes resulting from melanosome dispersion and aggregation are regulated by MSH. In teleost fishes it is unresolved whether both MCH and MSH control color changes or whether melanosome dispersion results solely from a cessation of MCH secretion.

The receptors through which MSH and MCH mediate their actions, are separate. The receptors for MSH require Ca^{2+} for the transduction process, whereas the receptors for MCH do not. The second messenger of MSH action is cAMP; the intracellular mechanism by which MCH induces melanosome aggregation remains undetermined.

ACKNOWLEDGMENTS

These studies were supported in part by grants from the Public Health Service (MH-27257, MH-30626, and AM-17420) from the National Science Foundation (PCM-883300, PCM-811220, PCM-770731, and PCM-810078), U.S.A., and from the Fundacao de Amparo a Pesquisa de Estrado de São Paulo (84/1967, 85/0718), Brasil.

REFERENCES

1. **Hadley, M. E.,** *Endocrinology,* 2nd ed., Prentice-Hall, Englewood Cliffs, N.J., 1988, 8.
2. **Bagnara, J. T. and Hadley, M. E.,** *Chromatophores and Color Change: Comparative Physiology of Animal Pigmentation,* Prentice-Hall, Englewood Cliffs, N.J., 1972.
3. **Hadley, M. E. and Quevedo, W. C., Jr.,** Vertebrate epidermal melanin unit, *Nature (London),* 209, 1334, 1966.
4. **Ide, H., Kawazoe, I., and Kawauchi, H.,** Fish melanin-concentrating hormone disperses melanin in amphibian melanophores, *Gen. Comp. Endocrinol.,* 58, 486, 1985.
5. **Abe, K., Butcher, R. W., Nicholson, W. E., Baird, C. E., Liddle, R. A., and Liddle, G. W.,** Adenosine 3′,5′-monophosphate (cyclic AMP) as the mediator of the actions of melanocyte stimulating hormone (MSH) and norepinephrine on the frog skin, *Endocrinology,* 84, 362, 1969.
6. **Abe, K., Robison, G. A., Liddle, G. W., Butcher, R. W., Nicholson, W. E., and Baird, C. E.,** Role of cyclic AMP in mediating the effects of MSH, norepinephrine and melatonin on frog skin color, *Endocrinology,* 85, 674, 1969.
7. **Bagnara, J. T., Taylor, J. D., and Hadley, M. E.,** The dermal chromatophore unit, *J. Cell. Biol.,* 38, 69, 1968.
8. **Hadley, M. E. and Bagnara, J. T.,** Integrated nature of chromatophore responses in the *in vitro* frog skin bioassay, *Endocrinology,* 84, 69, 1969.
9. **Bitensky, M. W. and Burstein, S. R.,** Effects of cyclic adenosine monophosphate and melanocyte-stimulating hormone on frog skin *in vitro, Nature (London),* 208, 1282, 1965.
10. **Goldman, J. M. and Hadley, M. E.,** The *beta* adrenergic receptor and cyclic 3′,5′-adenosine monophosphate: possible roles in the regulation of melanophore responses of the spadefoot toad, *Scaphiopus couchi, Gen. Comp. Endocrinol.,* 13, 151, 1969.
11. **Hadley, M. E. and Goldman, J. M.,** Physiological color change in reptiles, *Am. Zool.,* 9, 489, 1996.
12. **Fujii, R. and Oshima, N.,** Control of chromotophore movements in teleost fishes, *Zool. Sci.,* 3, 13, 1986.
13. **Pawelek, J.,** Factors regulating growth and pigmentation of melanoma cells, *J. Invest. Dermatol.,* 66, 201, 1976.
14. **Kreiner, P. W., Gold, C. J., Keirns, J. J., Brock, W. A., and Bitensky, M. W.,** MSH-sensitive adenyl cyclase in the Cloudman melanoma, *Yale J. Biol. Med.,* 46, 583, 1973.

15. **Bregman, M. D., Sawyer, T. K., Hadley, M. E., and Hruby, V. J.,** Adenosine and divalent cation effects on S-91 melanoma adenylate cyclase, *Arch. Biochem. Biophys.*, 200, 1, 1980.
16. **DeGraan, P. N. E., Gispen, W. H., and van de Veerdonk, F. C. G.,** α-Melanotropin-induced changes in protein phosphorylation in melanophores, *Mol. Cell. Endocrinol.*, 42, 119, 1985.
17. **DeGraan, P. N. E., Oestreacher, A. B., Zwiers, H., Gispen, W. H., and van de Veerdonk, F. C. G.,** Characterization of α-MSH-induced changes in the phosphorylation of a 53 kDa protein in *Xenopus* melanophores, *Mol. Cell. Endocrinol.*, 42, 127, 1985.
18. **Burnett, J. B., Birch, D. E., Fuller, B. B., and Hadley, M. E.,** Cyclic AMP-dependent protein kinase activity in Cloudman mouse melanoma cell cultures, in *Pigment Cell 1981*, Seiji, M. E., Ed., University of Tokyo Press, 1981, 355.
19. **Goldman, J. M. and Hadley, M. E.,** *In vitro* demonstration of adrenergic receptors controlling melanophore response of the lizard, *Anolis carolinensis*, *J. Pharmacol. Exp. Ther.*, 166, 1, 1969.
20. **Goldman, J. M. and Hadley, M. E.,** Evidence for separate receptors for melanophore stimulating hormone and catecholamine regulation of cyclic AMP in the control of melanophore responses, *Br. J. Pharmacol.*, 39, 160, 1970.
21. **Goldman, J. M. and Hadley, M. E.,** Sulfhydryl requirement for alpha adrenergic receptor activity and melanophore stimulating hormone (MSH) action on melanophores, *J. Pharmacol. Exp. Ther.*, 182, 93, 1972.
22. **Vesely, D. L. and Hadley, M. E.,** Calcium requirement for melanophore stimulating hormone action on melanophores, *Science*, 173, 923, 1971.
23. **Vesely, D. L. and Hadley, M. E.,** Receptor specific calcium requirement for melanophore stimulating hormone (MSH) control of melanophores, *Pigment Cell*, 3, 265, 1976.
24. **Vesely, D. L. and Hadley, M. E.,** Ionic requirements for melanophore stimulating hormone (MSH) action on melanophores, *Comp. Biochem. Phsyiol.*, 62A, 501, 1978.
25. **Hadley, M. E., Anderson, B., Heward, C. B., Sawyer, T. K., and Hruby, V. J.,** Calcium-dependent prolonged effects on melanophores of [4-norleucine, 7-D-phenylalanine] α-melanotropin, *Science*, 213, 1025, 1981.
26. **Hadley, M. E., Heward, C. B., Hruby, V. J., Sawyer, T. K., and Yang, Y. C. S.,** in *Peptides of the Pars Intermedia*, Ciba Symp. 81, Pitman Medical, London, 1981, 242.
27. **Hadley, M. E., Heward, C. B., Hruby, V. J., Sawyer, T. K., and Yang, Y. C. S.,** Hormone receptors of vertebrate pigment cells, in *Pigment Cell 1981*, Seijii, M. E., Ed., University of Tokyo Press, 1981, 323.
28. **Hadley, M. E. and Bagnara, J. T.,** Regulation of release and mechanism of action of MSH, in *Trends in Comparative Endocrinology*, Barrington, E. J. W., Ed., Thomas J. Griffiths Sons, New York, 1975, 81.
29. **Obika, M.,** Intracellular transport of pigment granules in fish chromatophores, *Zool. Sci.*, 3, 1, 1986.
30. **Akiyama, T. and Matsumoto, J.,** The blockade of pigment displacement in swordtail erythrophores by microinjection of antiactin antibody, *J. Exp. Zool.*, 227, 405, 1983.
31. **Green, L.,** Mechanism of movements of granules in melanocytes of *Fundulus heteroclitus*, *Proc. Natl. Acad. Sci. U.S.A.*, 59, 1179, 1968.
32. **Luby, K. L. and Porter, K. R.,** The control of pigment migration in isolated erythrophores of *Holocentrus ascensionis* (Osbeck). I. Energy requirements, *Cell*, 21, 13, 1980.
33. **Porter, K. R. and McNiven, M. A.,** The cytoplast: a unit structure in chromatophores, *Cell*, 29, 23, 1982.
34. **Wilkes, B. C., Hruby, V. J., Castrucci, A. M. de L., Sherbrooke, W. C., and Hadley, M. E.,** Synthesis of a cyclic melanotropic peptide exhibiting both melanin concentrating and dispersing activities, *Science*, 224, 111, 1984.
35. **Wilkes, B. C., Hruby, V. J., Sherbrooke, W. C., Castrucci, A. M. de L., and Hadley, M. E.,** Synthesis and biological actions of melanin concentrating hormone, *Biochem. Biophys. Res. Commun.*, 122, 613, 1984.
36. **Oshima, N., Kasukawa, H., Fujii, R., Wilkes, B. C., Hruby, V. J., Castrucci, A. M. de L., and Hadley, M. E.,** Melanin concentrating hormone (MCH) effects on teleost *(Chrysiptera cyanea)* melanophores, *J. Exp. Zool.*, 235, 175, 1985.
37. **Hruby, V. J., Wilkes, B. C., Cody, W. L., Sawyer, T. K., and Hadley, M. E.,** Melanotropins: structural, conformational and biological considerations in the development of superpotent and superprolonged analogs, in *Protein Report Review*, Vol. 3, Hearn, M. T. W., Ed., Marcel Dekker, New York, 1984, 1.
38. **Hadley, M. E., Wilkes, B. C., Zechel, C., Hruby, V. J., Castrucci, A. M. de L., Visconti, M. A., and Pozo-Alonso, M.,** Differential structural requirements for MSH and MCH activities of melanin concentracting hormone, *Life Sci.*, 40, 1139, 1986.
39. **Castrucci, A. M. de L., Hadley, M. E., and Hruby, A. J.,** Melanin concentrating hormone (MCH) exhibits both MSH and MCH activities on individual melanophores, *Life Sci.*, 40, 1845, 1987.
40. **Sawyer, T. K., Hruby, V. J., Hadley, M. E., and Engel, M. H.,** Melanocyte stimulating hormone: chemical nature and mechanism of action, *Am. Zool.*, 23, 529, 1983.

Chapter 3

THE ROLE OF PROTEIN PHOSPHORYLATION IN THE MECHANISM OF ACTION OF α-MSH

Pierre N. E. De Graan and Alex N. Eberle

TABLE OF CONTENTS

I. Introduction 28
  A. cAMP-Dependent Protein Phosphorylation 28
  B. Calcium-Dependent Protein Phosphorylation 29

II. Melanophore Systems 29
  A. Effects of α-MSH on Protein Phosphorylation 29
  B. Role of cAMP 32
  C. Calcium Requirement 35

III. Melanoma Cells 36
  A. Effects of α-MSH on Protein Phosphorylation 40
  B. Role of cAMP-Dependent Protein Kinases 40

IV. Other Target Tissues of α-MSH 40
  A. Xanthophores 40
  B. Lacrimal Gland 41
  C. Central Nervous System 41
  D. Adrenal Cells 42

V. Summary and Perspectives 42

Acknowledgments 43

References 43

I. INTRODUCTION

The peptide hormone α-melanotropin (α-MSH) is well known to stimulate pigment dispersion in dermal melanophores of lower vertebrates[1] and pigment formation in epidermal melanocytes[2] and melanoma cells.[3] α-MSH is also thought to play an important role in a number of other systems including fetal development,[4] axonal regeneration,[5,6] and modulation of neuronal activity and behavior.[7] Detailed reviews on the biological effects of melanotropins can be found in Volume II of this series.

Although the biological effects of α-MSH in most systems are well characterized, information on the molecular mechanism of action is limited and mainly derived from studies with amphibian melanophores and melanoma. The first step in α-MSH action on its target cell is thought to be specific hormone-receptor binding, resulting in the activation of a multistep mechanism eventually leading to diverse and complex biological responses (for review see Chapters 2, 4, and 11, this Volume). Receptor activation is thought to stimulate adenylate cyclase, thus increasing intracellular cAMP levels. Calcium has been shown to be required for MSH action and can be regarded as putative second messenger. Both second messengers are known to modulate the activity of specific protein kinases (PKs) and hence the degree of phosphorylation of substrate proteins in response to hormones in a variety of cell systems.[8] Substrate proteins may be, for example, enzymes, structural proteins, ion channels, receptor proteins, etc. and may be localized in any of the cellular compartments ranging from plasma membrane to nuclear proteins. cAMP has been shown to mediate cellular responses through cAMP-dependent PKs[9] a class of enzymes which is thought to be universally present in animal cells. An increase in intracellular calcium concentration may activate several types of PKs including calcium/calmodulin-dependent PKs[10,11] and the calcium/phospholipid-dependent PK known as protein kinase C (PKC).[12-14]

In this paper, the primary focus will be on the evidence for a role of cAMP-dependent PKs and their substrates in the mechanism of action of α-MSH in melanophores and melanoma. The evidence for an involvement of calcium/calmodulin-dependent PKs and PKC is very limited and only of a preliminary nature. Data from other target tissues of α-MSH pointing to a role of protein phosphorylation will be discussed only briefly.

A. cAMP-Dependent Protein Phosphorylation

The effects of the second messenger cAMP are mediated by cAMP-dependent protein kinases (cAMP-PKs) through the phosphorylation of a wide variety of substrate proteins. cAMP-PKs have been isolated from many different tissues and species and appear to vary little in structure or function.[10,11,15] The kinase has two subunits. The catalytic subunit (C) catalyzes the transfer of the γ-phosphate of the ATP to serine and/or threonine residues in the substrate proteins. The regulatory subunit (R) has the ability either to bind and inhibit the C or to bind cAMP. The holoenzyme consists of two Cs and two disulfide-linked Rs. The holoenzyme form of the kinase is inactive, and it appears that R inhibits the C by directly binding to the catalytic site of the enzyme. cAMP activates the enzyme by binding to the R. This binding causes dissociation of the holoenzyme into two free Cs, which possess cAMP-independent kinase activity, and one R dimer bound to cAMP. The activation of the enzyme by cAMP can be summarized as follows:

$$2\ \text{cAMP} + \text{R2C2} = \text{R2cAMP2} + 2\ \text{C}$$

(C = catalytic subunit, R = regulatory subunit, and R2C2 = holoenzyme).

Two types of cAMP-PKs were originally separated from rabbit skeletal muscle by DEAE-cellulose chromatography[16] and have since been isolated from many tissues. These forms, designated type I and type II, differ in their tissue distribution, their biochemical characteristics, and their immunological properties. The two types differ only in their Rs, designated R_I and R_{II}. The physiological significance of these two distinct forms of Rs has yet to be resolved.

B. Calcium-Dependent Protein Phosphorylation

Calcium has been shown to be an intracellular second messenger in the mechanism of action of many hormones and neurotransmitters.[17,18] In these signal-transduction systems receptor activation is thought to activate the enzyme phospholipase C. This phospholipase catalyzes the hydrolysis of a special class of membrane phospholipids, the polyphosphoinositides. Receptor-mediated hydrolysis of phosphatidylinositol 4,5-bisphosphate (PIP_2) results in the formation of diacylglycerol and inositol 1,4,5,-trisphosphate (IP_3). IP_3 is thought to induce a transient rise in cytosolic calcium; diacylglycerol stimulates the activity of PKC.[19] One of the mechanisms by which calcium exerts its intracellular actions is the regulation of protein phosphorylation. Two types of calcium-dependent PKs can be distinguished: calcium/calmodulin-dependent protein kinases (Cam-PKs) and the calcium/phospholipid dependent protein kinase, PKC.

Examples of the Cam-PKs are phosphorylase kinase, myosin light-chain kinase, and Cam-PK type I and II.[10,11] The activity of these kinases is regulated by the formation of a calcium/calmodulin complex, which binds to the enzyme. Thus, calmodulin, in effect, confers calcium sensitivity on the kinase, which, in the absence of calmodulin, is unresponsive to calcium. The four Cam-PKs mentioned above have all been purified and characterized. Phosphorylase kinase and myosin light-chain kinase exhibit narrow substrate specificity, whereas the Cam-PKs type I and II appear to phosphorylate a large number of substrate proteins. Cam-PK type II, which has been found in different tissues, is composed of subunits with molecular weights of 50,000 and 60,000 daltons, respectively. These subunits bind calmodulin and are autophosphorylated in a calcium/calmodulin-dependent manner.

The other type of calcium-dependent protein kinase is PKC.[12-14] Full activation of the enzyme has been shown to require the presence of calcium, phosphatidylserine, and diacylglycerol. Its activation by diacylglycerol can be mimicked by tumor-promoting phorbol esters.[20] One of the steps involved in PKC activation is the translocation of the kinase from the cytosol to the plasma membrane. Recently, the primary structure of PKC has been elucidated.[21-23] PKC has been implicated in many physiological processes, e.g., cell proliferation, exocytosis, and smooth muscle contraction.[14]

Although no data are available on MSH-induced changes in calcium- or calcium/calmodulin-dependent protein phosphorylation one cannot exclude the possibility that MSH may also mediate its effects through a transmembrane signal transduction system using calcium as second messenger. Recently, low concentrations of the related hormone adrenocorticotropin (ACTH) were shown to induce polyphosphoinositide hydrolysis in the adrenal.[24]

II. MELANOPHORE SYSTEMS

A. Effects of α-MSH on Protein Phosphorylation

As cultured melanophores are not available in large quantities, melanophores in the ventral tailfin of tadpoles of *Xenopus laevis*[25] were used to study the role of protein phosphorylation in the mechanism of action of α-MSH. For these studies the melanophore-rich caudal part of the ventral tailfins was excised and preincubated at room temperature for 60 min in assay medium to obtain full melanosome aggregation.[26] Subsequently, the tailfin pieces were incubated for 60 min with or without α-MSH to induce full pigment dispersion. Eight tailfins were homogenized in 80 $\mu\ell$ ice-cold phosphorylation buffer, and the homogenates were subjected to in vitro endogenous protein phosphorylation with [γ-^{32}P]-ATP.[26] Phosphorylated

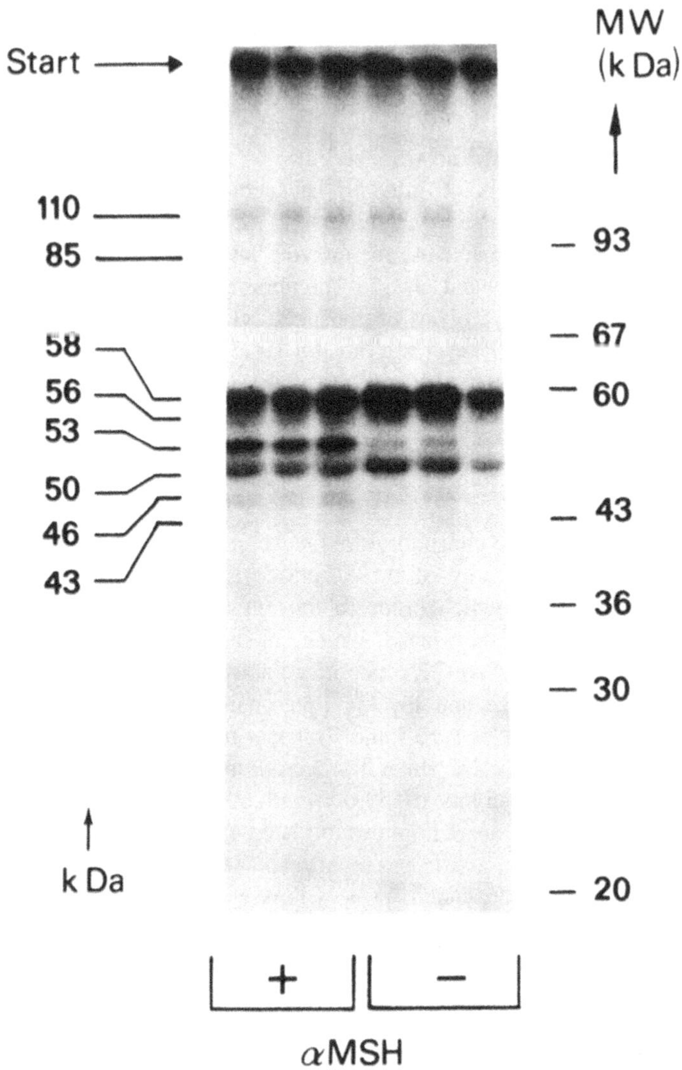

FIGURE 1. Effect of α-MSH treatment of intact melanophore-rich tailfins on [^{32}P]-incorporation into protein bands. Tailfins were incubated with or without 5 × 10^{-6} *M* α-MSH for 60 min, homogenized, and assayed for endogenous protein phosphorylation. Proteins were separated on 11% SDS-polyacrylamide gels and stained with Fast Green. After protein staining, gels were dried and subjected to autoradiography. The molecular weights of reference proteins are indicated on the left.

proteins were separated on 11% sodium dodecylsulfate (SDS)-polyacrylamide gel electrophoresis. The gels were dried and subjected to autoradiography to visualize [^{32}P]-incorporation into proteins. [^{32}P]-incorporation was quantified by densitometry or liquid scintillation counting.

Endogenous in vitro phosphorylation of tailfin homogenates resulted in [^{32}P]-incorporation into at least eight protein bands (Figure 1); the predominant bands have apparent molecular weights of 58, 53, and 50 kDa. A 60-min incubation of intact tailfins with 5 × 10^{-6} *M* α-MSH resulted in a fivefold increase in [^{32}P]-incorporation into a protein band with an apparent molecular weight of 53 kDa (Figure 1). [^{32}P]-incorporation into other phosphoproteins did not change significantly in response to α-MSH. Similar effects on 53-kDa phosphorylation were obtained after i.m. injection of α-MSH into white-background-adapted *Xenopus* tadpoles.[26] Manipulation of the endogenous blood-MSH levels by altering background adaptation conditions also resulted in changes in 53-kDa phosphorylation. Homogenates from tailfins of white-background adapted animals, which had been transferred to a black background 30 or 60 min prior to the tailfin excision, showed significantly more 53-kDa phosphorylation than those of tailfins from animals kept on a white background.[26]

To investigate the nature of the [^{32}P]-incorporation into the 53-kDa phosphoprotein, a number of control experiments were performed.

1. When α-ATP was used during the phosphorylation assay, no [^{32}P]-incorporation into proteins was observed.[27] This rules out the possibility of (non)-specific ATP binding to the 53-kDa band and indicates that the phosphorylation is a kinase-mediated reaction. The involvement of such an enzyme is further substantiated by the fact that boiling of the homogenate prior to the phosphorylation assay for 2 min completely abolished [^{32}P]-incorporation.
2. No loss of label could be detected during a 10-min incubation of a phosphorylated homogenate at 24°C. Thus, phosphatase activity in the homogenates seems to be negligible under the phosphorylation conditions.
3. Addition of a phosphorylated homogenate from α-MSH-treated tailfins back to a new extract during homogenization did not alter [^{32}P]-incorporation into the 53-kDa band. This indicates that the phosphoprotein is stable during processing.
4. Addition of α-MSH immediately after homogenization or 5 min prior to the phosphorylation assay did not affect 53-kDa phosphorylation, indicating that an intact melanophore system is required for the α-MSH effect.
5. To verify whether α-MSH-induced effects on 53-kDa phosphorylation are localized in melanophores, the melanophore-rich and melanophore-free parts of the tailfins were incubated separately with α-MSH and assayed for endogenous protein phosphorylation. A significant increase in 53-kDa phosphorylation was only found in the melanophore-rich part (Figure 2A). Moreover, the 53-kDa protein could be demonstrated in phosphorylated homogenates of α-MSH-treated primary cultured melanophores isolated from ventral tailfins of *X. laevis* tadpoles.[28]
6. The effect on 53-kDa phosphorylation appears to be specific for melanotropic peptides (Figure 2B). Incubation of tailfins with $ACTH_{1-24}$, an α-MSH-like peptide producing complete melanosome dispersion, also induced an increase in 53-kDa phosphorylation, whereas a structurally related peptide ($ACTH_{15-24}$) and an unrelated peptide (LH-RH), neither of which produced dispersion, were ineffective.

   The results of the control experiments described above indicate that the in vitro endogenous protein assay is a suitable assay to monitor MSH-induced changes in protein phosphorylation in intact melanophores.

The increase in 53-kDa phosphorylation is concentration dependent (Figure 3). A significant increase is detectable at 5×10^{-8} *M* α-MSH using a 15-min incubation.[29] After a 5-min incubation of the tailfins with α-MSH, the sensitivity is increased to 10^{-8} *M*. This concentration is about ten times higher than the concentration required to induce a visible pigment dispersion. This difference in sensitivity to α-MSH is most likely due to the limited sensitivity of the in vitro phosphorylation assay on total homogenates (melanophores constitute less than 1% of the tailfin cells).

Changes in 53-kDa phosphorylation are first detectable after a 5-min α-MSH treatment, a time which closely resembles the time required to produce a visible melanophore response (Figure 4). However, maximal effects on 53-kDa phosphorylation are found after 5 to 10 min, whereas maximal pigment dispersion is reached after 45 to 60 min. Thus, the effects on 53-kDa phosphorylation seem to precede maximal pigment dispersion. This suggests that

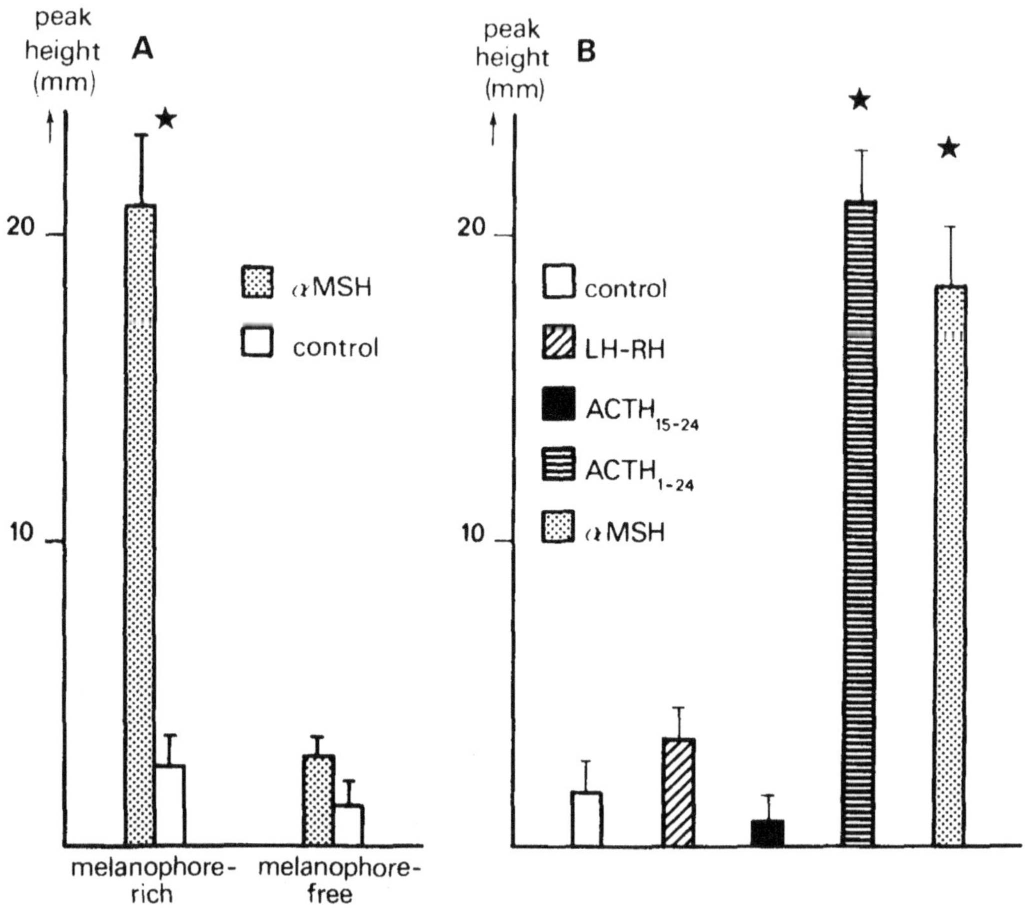

FIGURE 2. Concentration dependency of the α-MSH-induced effect on 53-kDa phosphorylation (n = 8; ★, significantly different from control, $p < 0.001$; bars indicate S.E.M.).

53-kDa phosphorylation may be related to the initiation of pigment dispersion rather than to the maintenance of the dispersed state of the melanosomes.

To determine the isoelectric point of the 53-kDa protein, a 10,000 × *g* supernatant of the tailfin homogenate was subjected to two-dimensional slab gel electrophoresis.[29] The 53-kDa band migrates as a single spot with an isoelectric point of 5.1 and was found to migrate in the acidic tail of purified dog brain β-tubulin (Figure 5B). The isoelectric point, the molecular weight, and the cytosolic localization of the 53-kDa protein indicate that the protein may be β-tubulin-like. A 53-kDa protein in the homogenate indeed cross-reacts on one-dimensional Western blots with monoclonal rat brain antitubulin IgM, but not with goat antidog brain tubulin antiserum, nor with a purified IgG fraction prepared from this serum (Figure 5A). Two-dimensional Western blotting was not performed. Although these data indicate that the 53-kDa protein may be tubulin-like, other identities also should be considered. As MSH has been shown to mediate its effect through cAMP, the 53 kDa may also be the autophosphorylated R of type II cAMP-PK (R_{II}). R_{II} has a molecular weight of 54,000 daltons, an isoelectric point of about 5 and a predominantly cytosolic localization. Detailed characterization studies will be required to unequivocally identify the 53-kDa protein.

B. Role of cAMP

In melanophores, α-MSH-receptor activation is thought to stimulate membrane-bound adenylate cyclase activity, thus increasing intracellular levels of cAMP.[30,31] High concen-

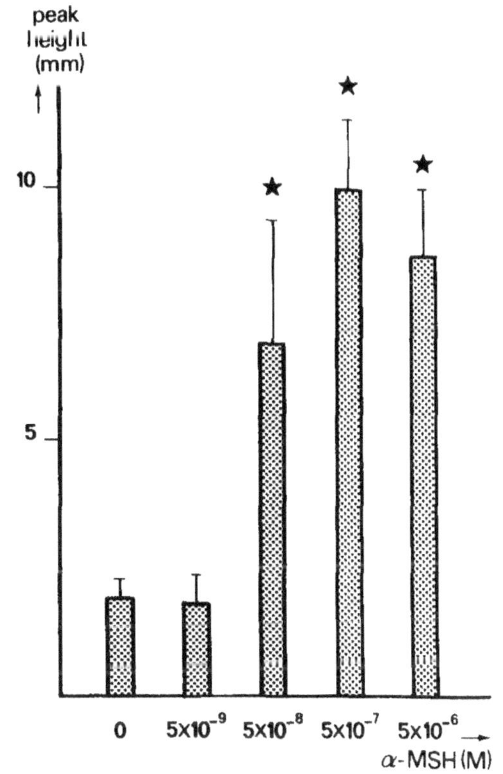

FIGURE 3. (A) Effect of α-MSH treatment (5 × 10^{-6} *M*) on 53-kDa phosphorylation in melanophore-rich and melanophore-free parts of the same tailfins (n = 8). (B) Effect of incubating tailfins with melanotropic and nonmelanotropic peptides on 53-kDa phosphorylation. The melanotropic peptides ($ACTH_{1-24}$ and α-MSH) and the nonmelanotropic ones ($ACTH_{15-24}$ and LH-RH) were all tested at 5 × 10^{-6} *M* (n = 6; ★, significantly different from control, $p < 0.001$; bars indicate S.E.M.).

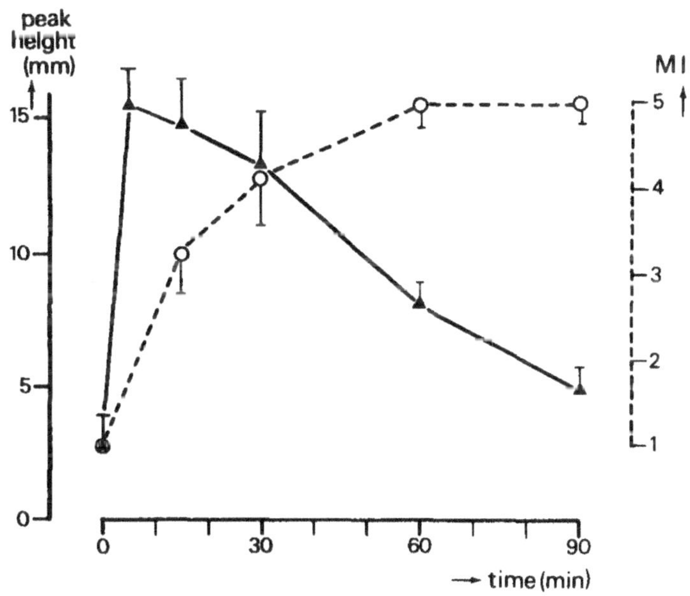

FIGURE 4. Time-response curve showing the effect of α-MSH (5 × 10^{-6} *M*) treatment of tailfins on 53-kDa phosphorylation (▲——▲) and pigment dispersion (○– – –○) (n = 6; MI, melanophore index; MI = 1 denotes complete pigment aggregation and MI = 5, full dispersion; bars indicate S.E.M.).

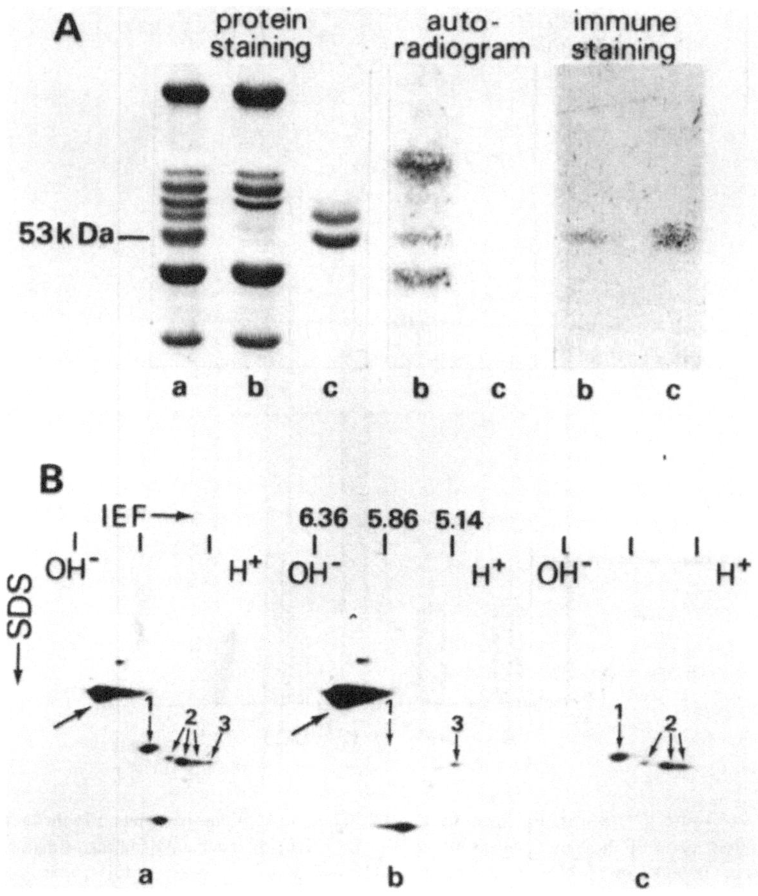

FIGURE 5. (A) Immunochemical characterization of the 53-kDa protein. Proteins (60 μg) of a homogenate from 5×10^{-6} *M* α-MSH-treated tailfins were separated on a 3-mm-thick polyacrylamide gel, and longitudinal sections were cut from the frozen gel. Gel slices were either stained with Fast Green for protein and subjected to autoradiography or analyzed for cross-reactivity with monoclonal rat brain antitubulin IgM, using the SDS-gel-immunoperoxidase method. Only the 40—80-kDa molecular weight region of the gel is shown. (Lane a, homogenate + 4 μg purified tubulin; lane b, homogenate; lane c, 4 μg purified tubulin). (B) Two-dimensional analysis of tailfin proteins (Fast Green staining). Phosphorylated proteins (100 μg) of the 10,000 × *g* supernatant of a homogenate from 5×10^{-6} *M* (15 min) α-MSH-treated tailfins were subjected to 2D polyacrylamide gel electrophoresis (a, supernatant + 8 μg purified tubulin; b, supernatant; c, 8 μg purified tubulin; 1, α-tubulin; 2, β-tubulin; 3, phosphorylated 53-kDa protein; ↑, bovine serum albumin from the incubation medium).

trations of extracellular cAMP and dibutytyl cAMP have been shown to be full agonists.[32,33] The role of cAMP in the mechanism of action of α-MSH is discussed in detail in Chapter 2 of this Volume. Melanophores in the ventral tailfin of *X. laevis* tadpoles show complete pigment dispersion in response to cAMP or dibutyryl cAMP.[34] Moreover, low concentrations of the diterpene, forskolin, a substance known to stimulate adenylate cyclase in many systems,[35] have been shown to stimulate pigment dispersion in *Xenopus* tailfin melanophores in a dose-dependent manner.[34]

Intracellular levels of cAMP are known to regulate cAMP-PK activity. Preliminary studies indicate that *Xenopus* tailfin homogenates contain cAMP-PK activity, but it is not possible to modulate in vitro endogenous 53-kDa phosphorylation in these homogenates with cAMP. (De Graan, unpublished). Incubation of tailfin pieces with forskolin induces a dose-dependent increase in the degree of 53-kDa phosphorylation assayed in vitro (Figure 6). Forskolin does not affect 53-kDa phosphorylation when added to tailfin homogenates in vitro. The dose

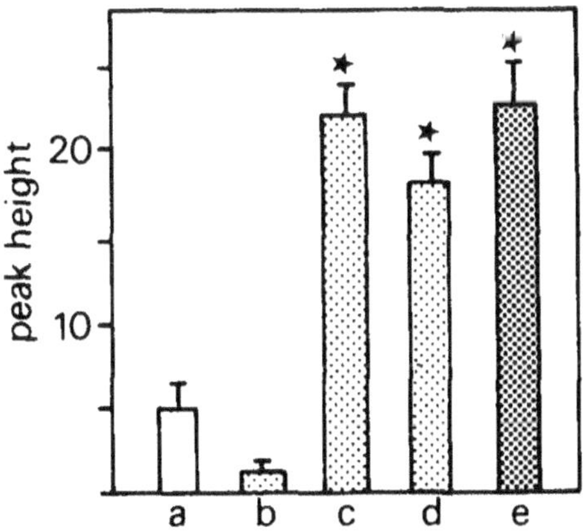

FIGURE 6. Effect of forskolin treatment of tailfins on in vitro endogenous phosphorylation of the 53-kDa melanophore protein (a, 0 *M*; b, 10^{-6} *M*; c, 10^{-5} *M*; d, 5×10^{-5} *M*; e, 5×10^{-6} *M* α-MSH; n = 12; ★, significantly different from control, $p < 0.001$; bars indicate S.E.M.; peak height is expressed in millimeters).

required for maximal stimulation of 53-kDa phosphorylation (10^{-5} *M*) is similar to the dose of forskolin required for maximal melanosome dispersion. Moreover, both forskolin and α-MSH stimulate 53-kDa phosphorylation to a maximum of four times the control level. These results support the hypothesis that forskolin and α-MSH stimulate melanophores through a common pathway involving cAMP-dependent 53-kDa phosphorylation.

C. Calcium Requirement

Extracellular calcium is well known to be required for the mechanism of action of α-MSH on melanophores of different species[33,36] (see also Chapter 2), including the melanophores in the ventral tailfin of *X. laevis* tadpoles.[37] Studies with synthetic α-MSH in calcium-containing and in calcium-free media, in the presence of calcium antagonists, and with ionophore A23187, revealed that extracellular calcium is required for one or more steps in MSH-signal transduction.[38] Applying photoreactive α-MSH derivatives to label MSH receptors covalently, at least two calcium sites could be shown; one associated with MSH-receptor binding and another with the subsequent generation of the biological effect.[39] The second calcium site is thought to precede the effect of cAMP, as cAMP-induced pigment dispersion does not require calcium.[36,38] In keeping with these findings, is the fact that forskolin-induced pigment dispersion of *Xenopus* tailfin melanophores does not require extracellular calcium.[34] As forskolin was shown to mimic the effect of α-MSH on 53-kDa phosphorylation (see above) it was of interest to know whether the α-MSH- and forskolin-induced increase in 53-kDa phosphorylation requires extracellular calcium.

Incubation of tailfins with 10^{-7} *M* α-MSH in the absence of extracellular calcium, does not result in melanosome dispersion, nor in stimulation of 53-kDa phosphorylation (Table 1). Forskolin, however, stimulates dispersion as well as 53-kDa phosphorylation in the absence of extracellular calcium. The inhibition of the MSH-induced stimulation by calcium deprivation is completely reversible; upon addition of 2 m*M* calcium to the extracellular medium, full redispersion as well as 53-kDa phosphorylation, is observed.[34] The calcium-free conditions per se do not affect basal 53-kDa phosphorylation levels. The difference in calcium requirement between forskolin and α-MSH probably resides in the mode of adenylate cyclase activation. Forskolin penetrates the plasma membrane and stimulates the cyclase directly,[35] whereas α-MSH stimulates the cyclase through a receptor-mediated pathway.

These results do not rule out other, for instance, intracellular calcium sites, because it is

Table 1
EFFECT OF CALCIUM DEPRIVATION ON α-MSH- AND FORSKOLIN-INDUCED STIMULATION OF 53-kDa PHOSPHORYLATION AND PIGMENT DISPERSION IN *XENOPUS* TAILFIN MELANOPHORES

Incubation conditions	[^{32}P]-incorporation (mm peak height)	Pigment dispersion (MI after 15 min)
Control + Ca^{2+}	2.2 ± 0.3 (14)	1.0
α-MSH + Ca^{2+}	12.0 ± 0.7[a] (20)	4.7
Forskolin + Ca^{2+}	10.5 ± 0.8[a] (20)	4.0
Control − Ca^{2+}	1.9 ± 0.3 (14)	1.0
α-MSH − Ca^{2+}	1.8 ± 0.3 (20)	1.0
Forskolin − Ca^{2+}	9.4 ± 0.8[a] (20)	4.8

Note: Calcium-free media contained 5 × 10^{-4} *M* EGTA. MI is melanophore index: MI = 1 denotes complete aggregation and MI = 5, full dispersion).

[a] Significantly different from control, $p < 0.001$; values are indicated ± SEM. Numbers in parentheses indicate number of experiments.

not known to what extent calcium deprivation conditions affect intracellular calcium levels. Preliminary studies indicate that in vitro endogenous 53-kDa phosphorylation is not calcium dependent. However, more purified melanophore preparations will be necessary to investigate a possible role of intracellular calcium in the regulation of 53-kDa phosphorylation.

III. MELANOMA CELLS

A. Effects of α-MSH on Protein Phosphorylation

α-MSH is known to stimulate pigment formation in melanoma cells.[3] Murine melanoma cells have been shown to contain specific MSH receptors,[40] which are linked to an adenylate cyclase system.[41] Within minutes after the addition of MSH to cultured murine melanoma cells, intracellular cAMP levels rise and eventually activate tyrosinase, the key enzyme for melanogenesis, and possibly other regulators of melanin biosynthesis.[42] The mechanism of tyrosinase activation is, however, poorly understood. Fuller and Viskochil[43] have shown that the activation of tyrosinase by MSH is dependent on *de novo* RNA and protein synthesis.

The effects of the intracellular second messenger cAMP are mediated through cAMP-PKs (see below). The information on substrate proteins on cAMP-PKs in melanoma cells is very limited. Only one report[44] describes α-MSH-induced changes in the degree of phosphorylation of a number of protein bands in cell lysates assayed for *post hoc* in vitro phosphorylation with [γ-^{32}P]-ATP. To obtain more direct information on the role of protein phosphorylation in the mechanism of action of α-MSH in melanoma cells, an in vivo phosphorylation assay for Cloudman S91 mouse melanoma cells has recently been developed.[44a]

In this assay, Cloudman S91 mouse melanoma cells (clone P3, 5 × 10^6 cells per milliliter) are loaded with [^{32}P]-labeled orthophosphate for 60 min to label endogenous ATP to a constant specific activity. Subsequently, the cells are incubated with or without α-MSH for 30 min. Incubated cells are boiled in a denaturing solution to stop the phosphorylation/dephosphorylation reactions, and melanoma cell proteins are separated by gradient SDS-polyacrylamide gel electrophoresis. The gels are stained with Fast Green, dried, and subjected to autoradiography. [^{32}P]-incorporation into phosphoproteins is quantified by densitometric scanning of the autoradiograms.

A 60-min incubation of melanoma cells with [^{32}P]-labeled orthophosphate resulted in the labeling of more than 30 phosphoproteins separated by one-dimensional SDS gel electro-

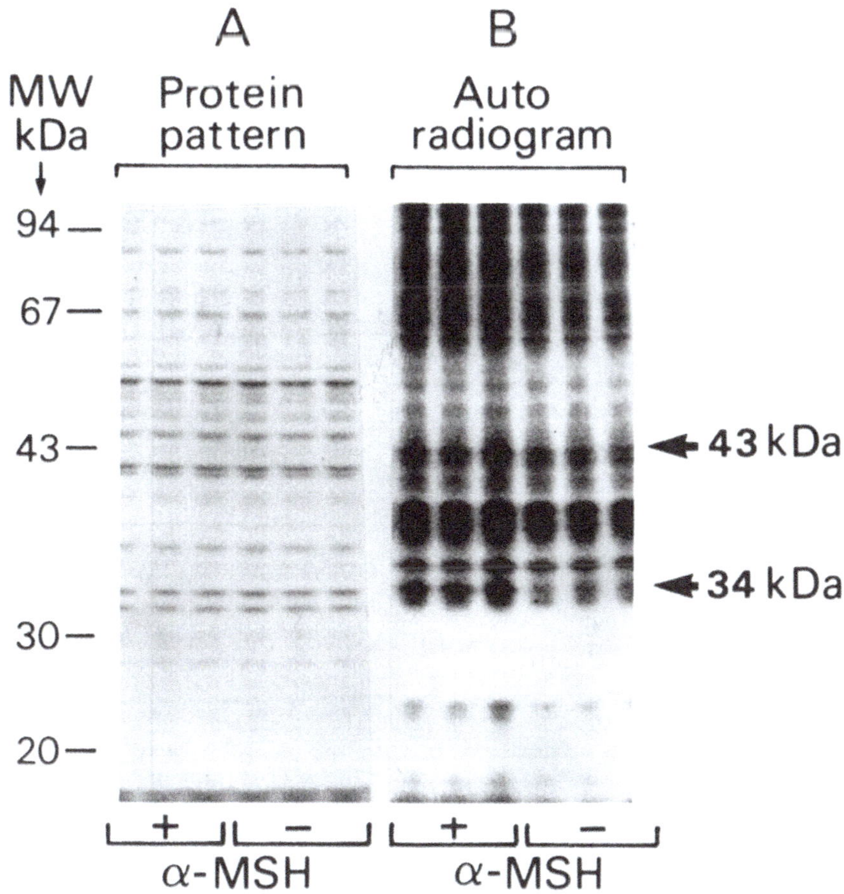

FIGURE 7. Protein pattern and autoradiogram after gel electrophoresis of solubilized Cloudman S91 cells which had been preincubated with ^{32}P, and then incubated with or without 10^{-6} *M* α MSH for 30 min. The positions of molecular weight marker proteins are indicated on the left.

phoresis (Figure 7). Incubation of the cells with 10^{-6} *M* α-MSH for 30 min did not affect the protein pattern, but resulted in an increase in [^{32}P]-incorporation in at least two protein bands with apparent molecular weights of 34 and 43 kDa respectively (Figure 7). Probably more protein bands are labeled and affected by the α-MSH treatment, but they are not resolved on the one-dimensional separation system. The [^{32}P]-labeling appears to be the result of a kinase-mediated reaction as the chemical linkage of the phosphate to the proteins was stable in acid, but labile in alkali. The effect of the hormone is not due to effects on the specific activity of endogenous [γ-^{32}P]-ATP pool as: (1) the specific activity of the ATP pool was not changed during the MSH treatment, (2) α-MSH did not change the overall [^{32}P]-incorporation into the proteins, and (3) the effect of α-MSH on 34- and 43-kDa phosphorylation was completely reversible, without affecting the [^{32}P]-incorporation into other protein bands (results not shown). Quantification of [^{32}P]-incorporation by densitometry of the autoradiograms showed the effects of α-MSH on 34- and 43-kDa phosphorylation to be concentration dependent (Figure 8). [^{32}P]-incorporation in the 43-kDa band was detectable at concentrations as low as 10^{-10} *M* α-MSH and reached a maximal stimulation at 10^{-8} *M* with a 156% stimulation compared to controls. [^{32}P]-incorporation into the 34-kDa band was significantly increased at 10^{-9} *M* α-MSH and continued to increase up to the highest α-MSH concentration tested (10^{-5} *M*) with a value of 250% of controls (Figure 8). Phos-

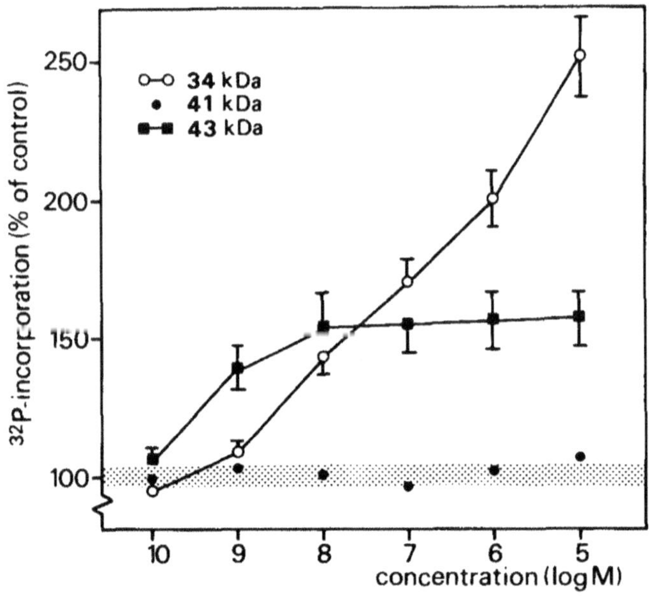

FIGURE 8. Dependence on α-MSH concentration of the [^{32}P-]incorporation into the 34-kDa, 41-kDa (control), and the 43-kDa protein band. Each gel track was scanned and peak height above background (PAB) determined. PAB values of MSH-stimulated melanoma cells were normalized against nonstimulated cells on the same gel. Each value is the mean of n = 6 determinations, except for 10^{-9} *M* (n = 5) and 10^{-5} *M* α-MSH (n = 4) and for the 41-kDa control band (n = 4). Bars indicate S.E.M. values.

phorylation of a control band (41 kDa) did not change in the MSH-concentration range tested.

The effect of α-MSH on 34- and 43-kDa phosphorylation appears to be specific for melanotropic peptides (Figure 9, only the effects on 34-kDa phosphorylation are shown). α-MSH was the most potent of the peptides tested. At a 10^{-6} *M* concentration, peptides stimulating tyrosinase and melanin production also induced an effect on 34- and 43-kDa phosphorylation. $ACTH_{11-24}$, β-endorphin, and VIP, all nonmelanotropic peptides, were without effect on phosphorylation.

The onset of the α-MSH-induced changes in 34-kDa phosphorylation is much more rapid than that in 43-kDa phosphorylation (Figure 10). After a 5-min incubation with α-MSH, the 34-kDa phosphorylation was almost maximal whereas 43-kDa phosphorylation did not significantly differ from nonstimulated controls. After a 30- to 60-min incubation with α-MSH there as a decline in 34-kDa phosphorylation, whereas the 43-kDa phosphorylation was still increasing. The time-course of the MSH-induced effects on the 34-kDa protein resembles that of cAMP production.[43] These findings are not surprising in view of preliminary results on the subcellular localization of the two proteins, which indicate that the 34-kDa protein is enriched in the membrane fractions, whereas the 43-kDa protein is predominantly found in the mitochondrial/melanosomal fraction. (Eberle and De Graan, unpublished results).

The 34-kDa protein appears to be the more interesting one to study in relation to receptor activation and transmembrane signal transduction. The fact that [^{32}P]-incorporation did not reach a plateau at α-MSH concentrations inducing maximal tyrosinase stimulation (EC_{50}: 2.5×10^{-9} *M* and EC_{100}: 10^{-7} *M*) may either point to an effect not directly related to tyrosinase stimulation or to the possibility that only a small change in the degree of 34-kDa

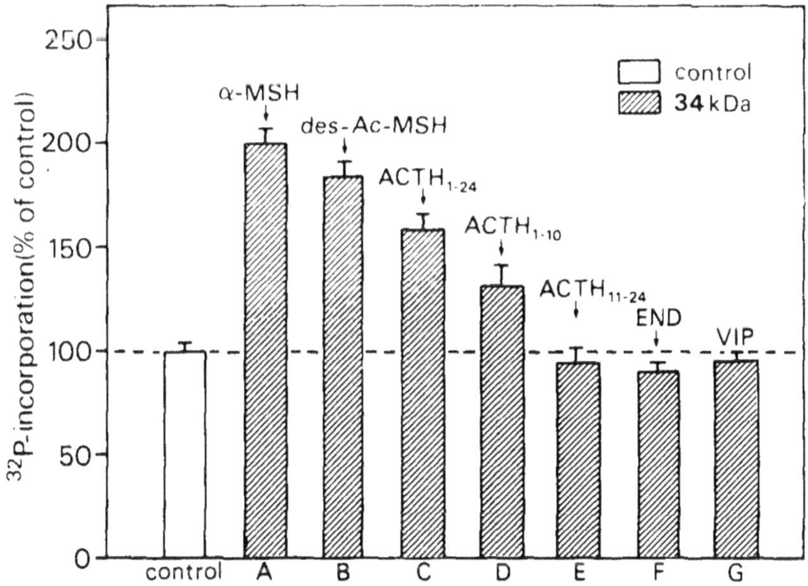

FIGURE 9. Effect of 10^{-6} *M* α-MSH (A), des-acetyl-α-MSH (B), $ACTH_{1-24}$ (C), $ACTH_{1-10}$ (D), $ACTH_{11-24}$ (E), β-endorphin (F), and vasointestinal peptide (G) on [^{32}P]-incorporation into the 34-kDa protein band. Each value is the mean of n = 12 (A), n = 6 (C), or n = 4 (all others). Bars indicate S.E.M. values.

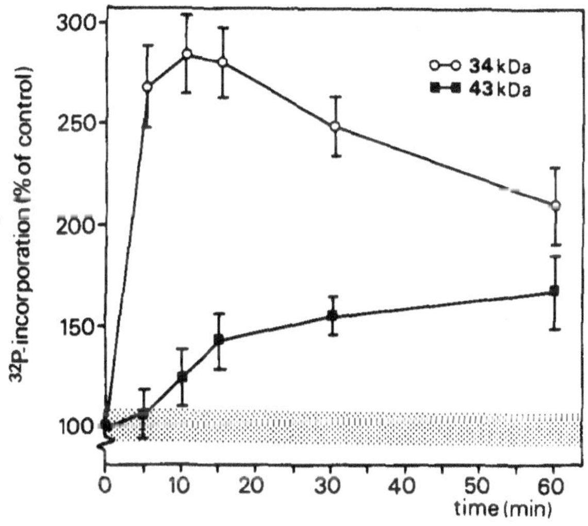

FIGURE 10. Time-course of [^{32}P]-incorporation into the 34- and 43-kDa protein bands induced by treatment of Cloudman S91 melanoma cells with 10^{-6} *M* α-MSH.

phosphorylation is required to elicit the effect. This is reminiscent of the concept of intracellular regulation by cAMP, where in many biological systems a fractional change was thought to be sufficient for maximal stimulation. Indeed maximal stimulation of adenylate cyclase in Cloudman S91 cell membranes was found at 10^{-5} *M* and an EC_{50} at concentrations similar to those required for inducing changes in 34-kDa phosphorylation.

Table 2
EFFECT OF db-cAMP, FORSKOLIN, AND IBMX ^{32}P-INCORPORATION INTO THE 34-kDa BAND (30-MIN ASSAYS) IN CLOUDMAN S91 MOUSE MELANOMA CELLS

Treatment	PAB	% of control
Control	6.7 ± 0.3 (9)	100 ± 5
α-MSH (10^{-6} *M*)	13.3 ± 0.5 (5)	199 ± 8[a]
db-cAMP (10^{-3} *M*) + IBMX (10^{-4} *M*)	15.2 ± 1.4 (4)	288 ± 20[a]
Forskolin (10^{-4} *M*) + IBMX (10^{-4} *M*)	13.0 ± 0.6 (7)	195 ± 9[a]
IBMX (10^{-4} *M*)	9.2 ± 0.5 (5)	138 ± 8[a]

Note: PAB, peak height above background.

[a] Significantly different from control, p <0.001; numbers in parentheses indicate number of experiments. Values are indicated ± S.E.M.

B. Role of cAMP-dependent Protein Kinases

In Cloudman S91 melanoma cells, cAMP-PK activity has been demonstrated.[45,46] Both type I and type II cAMP-PKs appear to be present, and the time course of cAMP-PK activation by α-MSH resembles that of MSH-induced generation of cAMP.[47] To our knowledge no information is available on the endogenous substrate proteins of cAMP-PKs in melanoma cells.

To investigate the possibility that the 34- and 43-kDa phosphoproteins which are affected by α-MSH treatment (see above) are cAMP-PK substrates, melanoma cells were incubated with substances known to elevate intracellular levels of cAMP and known for their ability to mimic the effects of α-MSH (Table 2). Dibutyryl-cAMP in the presence of the phosphodiesterase inhibitor isobutylmethylxanthine (IBMX) increased 34-kDa phosphorylation by 228% of the controls. Forskolin stimulated 34-kDa phosphorylation in the presence of IBMX by 195% of the controls (Table 2). Although cAMP-PK activity in Cloudman S91 cells was found to be increased by MSH,[47] the possibility that the effects of α-MSH on 34-kDa phosphorylation are (also) mediated by other kinases cannot be ruled out. For instance, ACTH and related peptides have been shown to inhibit the activity of PKC in rat brain synaptic plasma membranes.[48-50] In the adrenal gland, low concentrations of ACTH have been reported to stimulate polyphosphoinositide hydrolysis[24] (see below); this hydrolysis is thought to result in the formation of diacylglycerol, an intracellular messenger which has been shown to activate PKC[14] and inositol 1,4,5-trisphosphate,[18] an intracellular messenger which is thought to increase cytosolic calcium levels, thus activating Cam PKs.

IV. OTHER TARGET TISSUES OF α-MSH

A. Xanthophores

α-MSH, ACTH, and cAMP, all of which induce the dispersion of the carotenoid droplets in goldfish xanthophores, have been shown to stimulate the incorporation of phosphate into a 57-kDa protein found in association with the carotenoid droplets.[51] Using purified xanthophores prelabeled with $^{32}P_i$, Lynch et al.[52] recently showed that ACTH- and cAMP-induced droplet dispersion is accompanied by the phosphorylation of a pair of 57-kDa proteins called p57. Nearly all of this p57 protein is found to be associated with the carotenoid droplets. The minimal effective dose of ACTH or cAMP required to induce full pigment dispersion also fully stimulated the p57 phosphorylation. After two-dimensional gel electrophoresis the p57 proteins resolve into multiple charge variants which were shown to relate to multiple phosphorylation sites.[52] The increase in p57 phosphorylation is detectable within minutes after stimulation of the cells and appears to reach its maximum during the early phases of pigment dispersion. Withdrawal of ACTH induces rapid dephosphorylation of p57 which is complete before the pigment organelles have completely reaggregated.[52] Lynch and colleagues.[53] also provided evidence that the ACTH-induced increase in p57 phosphorylation is mediated through a cAMP-PK in the cytosol. They suggest that the activity of the cAMP-PK is important for the regulation of the intracellular movement of the pigment organelles of the xanthophore and that the phosphorylation and dephosphorylation of p57 in the xanthophore are crucial to the mechanism of pigment organelle movement.

B. Lacrimal Gland

α-MSH and ACTH have been shown to induce exocytosis in the rat lacrimal gland through a cAMP-dependent pathway.[54] Moreover, these peptides appear to induce an increase in the degree of in vivo phosphorylation of three gland proteins; a 35-kDa protein, called exocytosis-coupled protein (EC protein); a 26-kDa protein (protein II); and a 20-kDa protein (protein III).[55] The effect of α-MSH and ACTH on the phosphorylation of these three proteins could be mimicked by dibutyryl cAMP and IBMX.[54] The identity of the three proteins and that of the kinase involved is as yet unknown. However, all three proteins are also found in the parotid gland.[54] The EC protein is of particular interest as its phosphorylation also seems to be increased in response to secretagogues known to use calcium as second messenger.[56] The EC protein in the lacrimal gland has three properties in common with the 34-kDa described in melanoma cells (see above); (1) molecular weight, (2) MSH-induced cAMP-mediated phosphorylation, and (3) predominant localization in membrane-enriched fractions. It will be of interest to know whether there are more similarities between these proteins.

C. Central Nervous System

ACTH and MSH-like peptides have a large variety of effects on the central nervous system (CNS) and have been shown to be endogenous neuropeptides.[7] They are endogenous peptides derived from a large precursor molecular termed pro opiomelanocortin (POMC), from which other neuropeptides (e.g., β-LPH, endorphins) can also be enzymatically generated.[57,58] Cells synthesizing this precursor are not only present in the anterior and intermediate lobes of the pituitary, but are also found in the CNS. These cells are located in the ventral hypothalamus (*n. arcuatus*) and constitute a neuronal system with diffuse projections throughout the brain.[59] The neuropeptides produced by and released from neurons may act as neurotransmitters or neuromodulators.

ACTH and related neuropeptides modulate chemical neurotransmission in certain synapses in the mammalian brain by influencing at least two pathways of receptor-mediated transmembrane signal transduction. Firstly, ACTH affects the activity of adenylate cyclase in a number of systems, possibly through an effect of the peptide on the phosphorylation of one or more membrane proteins. This topic has recently been reviewed elsewhere.[60] Secondly, ACTH and MSH-like peptides have been shown to inhibit the phosphorylation of a neuron-specific membrane protein called B-50.[61] B-50 is a substrate of the PKC. The effect of the neuropeptides on the degree of B-50 phosphorylation is elicited by an inhibition of PKC activity.

The degree of B-50 phosphorylation is thought to be an important factor in the regulation of the activity of the enzyme, phosphatidylinositol 4-phosphate kinase (PIP kinase), the rate-limiting enzyme in the synthesis of phosphatidylinositol 4,5-bisphosphate (PIP_2). Receptor-mediated PIP_2 hydrolysis is known to result in the formation of the second messengers diacylglycerol and inositol 1,4,5-bisphosphate.[17,18] Since diacylglycerol stimulates PKC, thus increasing B-50 phosphorylation, the B-50 protein is thought to be part of a negative feedback loop in receptor-mediated hydrolysis of polyphosphoinositides. The inhibition of B-50 phosphorylation by ACTH and MSH-like peptides would temporarily attenuate the feedback regulation, thus rendering more PIP_2 available for receptor-mediated hydrolysis. The precise mechanism by which the neuropeptides inhibit PKC activity has not yet been resolved. However, these data show that a role of ACTH and MSH-like peptides in receptor-mediated hydrolysis of polyphosphoinositides has to be considered. This notion is further substantiated by data on the effects of ACTH on adrenal cells.

D. Adrenal Cells

Although it has generally been accepted that ACTH operated through the second messenger cAMP in the adrenal,[62,63] calcium also seems to be important.[64] Recently, it has been shown that low concentrations of ACTH induce an increase in inositol phosphate labeling and intracellular calcium in dissociated rat adrenal cells;[24] 10^{-12} *M* ACTH provoked a maximal increase in [^{3}H]-IP, IP_2, and IP_3 labeling of approximately 50, 150, and 60%, respectively. This ACTH concentration also produced a maximal increase in the cytosolic calcium concentration as measured with Quin-2.[24] At ACTH concentrations of 10^{-10} *M* and above, no effects were detectable on [^{3}H]-labeled inositol phosphates, nor on intracellular calcium levels. Stimulation of cAMP production was increased at ACTH concentrations as low as 10^{-12} *M*, but was more pronounced at 10^{-9} and 10^{-8} *M*. In the same adrenal cell system, corticosterone production was shown to increase linearly between 10^{-14} and 10^{-9} *M* $ACTH_{1-24}$. Thus, it seems that in a mixed adrenal cell population low concentrations of ACTH can activate both the cAMP and the IP_3-calcium intracellular signaling systems.

V. SUMMARY AND PERSPECTIVES

The first step toward defining a role of protein phosphorylation in the mechanism of action of α-MSH has been taken. In at least three target tissues of α-MSH it has now been shown that α-MSH induces changes in the degree of phosphorylation of a number of proteins. In *Xenopus* melanophores,[26,29] Cloudman S91 mouse melanoma cells,[42,44a] and goldfish xanthophores,[52,53] MSH-induced changes in protein phosphorylation appear to be rapid and reversible. The time-courses of these effects on phosphorylation suggest that phosphorylation is involved in the initiation of cellular response to α-MSH, such as pigment dispersion and synthesis. However, it is too early to assign a specific role of protein phosphorylation to a particular cellular process within pigment cells.

It will be essential to characterize and identify the major physiological substrate phosphoproteins affected by α-MSH treatment. Physiological substrates will have to be studied in intact cell systems using $^{32}P_i$ to label the endogenous ATP pool. *Post hoc* in vitro phosphorylation assays are usually less suitable, as they may reveal affects of α-MSH on substrates which are not phosphorylated by the same kinase in vivo. Two-dimensional separation systems as well as subcellular fractionation will be required to provide sufficient resolution to detect the physiological substrates. Substrates may range from the MSH receptor and components of the transmembrane signal transduction system to proteins involved in pigment organelle translocation and pigment synthesis. Also phosphoproteins involved in general melanocyte metabolism should be considered, as the physiological responses to α-MSH will require energy and may affect cell proliferation.

The kinase mediating the effects of α-MSH will most likely be a cAMP-PK. To provide direct evidence for a role of such a kinase in α-MSH-induced physiological responses, such as pigment dispersion and synthesis, one could try to manipulate the phosphorylation reaction. This may be achieved by microinjection of the kinase or its catalytic subunit (C), thus increasing the degree of phosphorylation of the substrate proteins. Alternatively, injection of antibodies against the kinase or the substrate protein may reduce the phosphate incorporation in the substrates. If these manipulations indeed affect the physiological response induced by α-MSH concomitantly with the phosphorylation reaction, one could conclude that the phosphorylation is an essential step in the mechanism of action of α-MSH.

Although cAMP-PKs are thought to mediate most of the effects of α-MSH, the involvement of other kinases should also be considered. The calcium requirement for α-MSH action suggests that calcium may be a second messenger and that PKC and Cam-PKs could be involved in the mechanism of α-MSH action. A precedent for such a complex response to receptor activation has been described in the adrenal.[24] In this system, ACTH has been shown to activate both cAMP and calcium-generating signal transduction pathways. It will be of great interest to study the possibility of such a dual activation mechanism in melanocytes. Possibly the two transduction systems serve to discriminate between the short- and long-term effects of α-MSH.

ACKNOWLEDGMENTS

We thank Mrs. L. Claessens for typing the manuscript and Mr. E. Kluis for preparing the figures. Part of this work was supported by the Swiss National Science Foundation and the Roche Research Foundation.

REFERENCES

1. **Bagnara, J. T. and Hadley, M. E.,** *Chromatophores and Color Change*, Prentice-Hall, Englewood Cliffs, N.J., 1973.
2. **Snell, R. S.,** Effect of α-MSH of the pituitary on mammalian epidermal melanocytes, *J. Invest. Dermatol.*, 42, 337, 1964.
3. **Wong, G. and Pawelek, J.,** Control of phenotypic expression of cultured melanoma cells by melanocyte stimulating hormone, *Nature (London) New Biol.*, 241, 213, 1973.
4. **Swaab, D. F. and Martin, J. T.,** Functions of α-melanotropin and other opiomelanocortin peptides in labour, intrauterine growth and brain development, in *Peptides of the Pars Intermedia*, Evered, D. and Lawrenson, G., Eds., Pitman Medical, London, 1981, 196.
5. **Strand, F. L. and Smith, C. M.,** LPH, ACTH, MSH and motor systems, *Pharmacol. Ther.*, 11, 509, 1980.
6. **Edwards, P. M. and Gispen, W. H.,** Melanocortin peptides and normal plasticity, in *Senile Dementia of the Alzheimer's Type: Early Diagnosis, Neuropathology and Animal Models*, Gispen, W. H. and Traber, J., Eds., Springer-Verlag, Heidelberg, 1985, 231.
7. **De Wied, D. and Jolles, J.,** Neuropeptides derived from proopiocortin: behavioral, physiological, and neurochemical effects, *Physiol. Rev.*, 62, 976, 1982.
8. **Rodnight, R.,** Cyclic nucleotides as second messengers in synaptic transmission, *Intl. Rev. Biochem.*, 26, 1, 1979.
9. **Greengard, P.,** Phosphorylated proteins as physiological effectors: protein phosphorylation may be a final common pathway for many regulatory agents, *Science*, 199, 146, 1979.
10. **Nestler, E. J. and Greengard, P.,** *Protein Phosphorylation in the Nervous System*, John Wiley & Sons, New York, 1984.
11. **Cohen, P.,** The role of protein phosphorylation in the hormonal control of enzyme activity, *Eur. J. Biochem.*, 151, 439, 1985.
12. **Kikkawa, U., Takai, Y., Minakuchi, R., Inohara, S., and Nishizuka, Y.,** *J. Biol. Chem.*, 257, 13341, 1982.
13. **Nishizuka, Y.,** The role of protein kinase C in cell surface signal transduction and tumor promotion, *Nature (London)*, 308, 693, 1984.
14. **Nishizuka, Y.,** Studies and perspectives of protein kinase C, *Science*, 233, 305, 1986.
15. **Schramm, M. and Selinger, Z.,** Message transmission: receptor controlled adenylate cyclase system, *Science*, 225, 1350, 1984.

16. **Reimann, E. M., Walsh, D. A., and Krebs, E. G.,** Purification and properties of rabbit skeletal muscle adenosine 3′,5′-monophosphate-dependent protein kinases, *J. Biol. Chem.*, 246, 1986, 1971.
17. **Abdel-Latif, A. A.,** Calcium-mobilizing receptors, polyphosphoinositides, and the generation of second messengers, *Pharmacol. Rev.*, 38, 227, 1986.
18. **Berridge, M. J. and Irvine, R. F.,** Inositol trisphosphate, a novel second messenger in cellular signal transduction, *Nature (London)*, 312, 315, 1984.
19. **Bell, R. M.,** Protein kinase C activation by diacylglycerol second messengers, *Cell*, 45, 631, 1986.
20. **Castagna, M., Takai, Y., Kaibuchi, K., Sano, K., Kikkawa, V., and Nishizuka, Y.,** Direct activation of calcium-activated, phospholipid-dependent protein kinase by tumor-promoting phorbol esters, *J. Biol. Chem.*, 257, 7847, 1982.
21. **Parker, P. J., Coussens, L., Totty, N., Rhee, L., Young, S., Chen, E., Stabel, S., Waterfield, M. D., and Ullrich, A.,** The complete primary structure of protein kinase C — the major phorbol ester receptor, *Science*, 233, 853, 1986.
22. **Coussens, L., Parker, P. J., Rhee, L., Yang-Feng, T. L., Chen, E., Waterfield, M. D., Francke, U., and Ullrich, A.,** Multiple, distinct forms of bovine and human protein kinase C suggest diversity in cellular signaling pathways, *Science* , 233, 859, 1986.
23. **Ono, Y., Kurokawa, T., Kawahara, K., Nishimura, O., Marumoto, R., Igarashi, K., Sugino, Y., Kikkawa, U., Ogita, K., and Nishizuka, Y.,** Cloning of rat brain protein kianse C complementary DNA, *FEBS Lett.*, 203, 111, 1986.
24. **Farese, R. V., Rosic, N., Babischkin, J., Farese, M. G., Foster, R., and Davis, J. S.,** Dual activation of the inositol-triphosphate-calcium and cyclic nucleotide intracellular signaling systems by adrenocorticotropin in rat adrenal cells, *Biochem. Biophys. Res. Commun.*, 135, 742, 1986.
25. **De Graan, P. N. E., Molenaar, R., and van de Veerdonk, F. C. G.,** A new *in vitro* melanophore bioassay for MSH using tail-fins of *Xenopus* tadpoles, *Mol. Cell. Endocrinol.*, 32, 271, 1983.
26. **De Graan, P. N. E., Gispen, W. H., and van de Veerdonk, F. C. G.,** α-Melanotropin-induced changes in protein phosphorylation in melanophores, *Mol. Cell. Endocrinol.*, 42, 119, 1985.
27. **De Graan, P. N. E., Zwiers, H., Oestreicher, A. B., Sweep, C. G. J., Van de Water, C., Van de Kamp, A. J., Gispen, W. H., and van de Veerdonk, F. C. G.** , *In vitro* phosphorylation of melanophore proteins: changes induced by α-MSH, in *Pigment Cell 1985, Proc. 12th Int. Pigment Cell Conf.*, Bagnara, S. N., Klaus, E., and Schartl, M., Eds., University of Tokyo Press, 1985, 197.
28. **Seldenrijk, R., Hup, D. R. W., De Graan, P. N. E., and van de Veerdonk, F. C. G.,** Morphological and physiological aspects of melanophores in pituitary culture from tadpoles of *Xenopus laevis, Cell Tissue Res.*, 198, 397, 1979.
29. **De Graan, P. N. E., Oestreicher, A. B., Zwiers, H., Gispen, W. H., and van de Veerdonk, F. C. G.,** Characterization of α-MSH-induced changes in the phosphorylation of a 53 kDa protein in *Xenopus* melanophores, *Mol. Cell. Endocrinol.*, 42, 127, 1985.
30. **Abe, K., Butcher, R. W., Nicholson, W. E., Baird, C. E., Liddle, R. A., and Liddle, G. W.,** Adenosine 3′,5′-monophosphate (cyclic AMP) as the mediator of the actions of melanocyte stimulating hormone (MSH) and norepinephrine on the frog skin, *Endocrinology*, 84, 362, 1969.
31. **Van de Veerdonk, F. C. G. and Konijn, T. M.,** The role of adenosin 3′,5′-cyclic monophosphate and catecholamines in the pigment migration process in *Xenopus laevis, Acta Endocrinol.*, 64, 364, 1970.
32. **Bitensky, M. W. and Burstein, S. R.,** Effects of cyclic adenosin monophosphate and MSH on frog skin in vitro, *Nature (London)*, 208, 1282, 1965.
33. **Vesely, D. L. and Hadley, M. E.,** Ionic requirements for melanophore stimulating hormone (MSH) action on melanophores, *Comp. Biochem. Physiol.*, 62A, 501, 1979.
34. **De Graan, P. N. E., Van de Kamp, A. J., Hup, D. R. W., Gispen, W. H., and van de Veerdonk, F. C. G.,** Calcium requirement for α-MSH action on melanophores, *J. Receptor Res.*, 4, 521, 1984.
35. **Seamon, K. B. and Daly, J. W.,** Forskolin, cAMP and cellular physiology, *Trends Pharmacol. Sci.*, 4, 120, 1983.
36. **Vesely, D. L. and Hadley, M. E.,** Calcium requirement for melanophore-stimulating hormone action on melanophores, *Science*, 173, 923, 1971.
37. **De Graan, P. N. E., Eberle, A. N., and van de Veerdonk, F. C. G.,** Photoaffinity labelling of MSH receptors reveals a dual role of calcium in melanophore stimulation, *FEBS Lett.*, 129, 113, 1981.
38. **De Graan, P. N. E., Van Dorp, C. J. M. M., and van de Veerdonk, F. C. G.,** Calcium requirement for α-MSH on tail-fin melanophores of *Xenopus* tadpoles, *Mol. Cell. Endocrinol.*, 26, 315, 1982.
39. **De Graan, P. N. E., Eberle, A. N., and van de Veerdonk, F. C. G.,** Calcium sites in MSH stimulation of *Xenopus* melanophores. Studies with photoreactive α-MSH, *Mol. Cell. Endocrinol.*, 26, 327, 1982.
40. **Eberle, A. N.,** Structure and chemistry of the peptide hormones of the intermediate lobe, in *Peptides of the Pars Intermedia*, Evered, D. and Lawrenson, G., Eds., Pitman Medical, London, 1981, 13.
41. **Bitensky, M. W. and Demopoulos, H. B.,** *J. Invest. Dermatol.*, 54, 83, 1970.
42. **Pawelek, J., Korner, A., Bergstrom, A., and Bologna, J.,** New regulators of melanin biosynthesis and the autodestruction of melanoma cells, *Nature (London)*, 286, 617, 1980.

43. **Fuller, B. B. and Viskochil, D. H.,** The role of RNA and protein synthesis in mediating the action of MSH on mouse melanoma cells, *Life Sci.*, 24, 2405, 1979.
44. **Pawelek, J., Fleischmann, R., McLane, J., Guillette, B., Emanuel, J., Korner, A., Bergstron, A., and Murray, M.,** Studies on growth and pigmentation of Cloudman S91 melanoma cells, in *Pigment Cell 1985, Proc. 12th Intl. Pigment Cell Conf.*, Bagnara, J., Klaus, S. N., Paul, E., and Schartl, M., Eds., University of Tokyo Press, 1985, 521.

44a. **De Graan, P. N. E., Brussaard, A. B., Gamboni, G., Girard, J., and Eberle, A. N.,** α-MSH-induced changes in protein phosphorylation of Cloudman S91 mouse melanoma cells, *Mol. Cell. Endocr.*, 57, 87, 1987.

45. **Korner, A. and Pawelek, J.,** Activation of melanoma tyrosinase by a cyclic AMP-dependent protein kinase in a cell-free system, *Nature (London)*, 267, 444, 1977.
46. **Niles, R. M. and Logue, M. P.,** *J. Supramol. Struct.*, 11, 251, 1979.
47. **Burnett, J. B., Bireh, D. E., Fuller, B. B., and Hadley, M. E.,** Cyclic-AMP dependent protein kinase activity in Cloudman mouse melanoma cell cultures, in *Pigment Cell*, Seiji, M.,Ed., University of Tokyo Press, 1981, 335.
48. **Zwiers, H., Veldhuis, D., Schotman, P., and Gispen, W. H.,** ACTH, cyclic nucelotides and brain protein phosphorylation in vitro, *Neurochem. Res.*, 1, 669, 1976.
49. **Zwiers, H., Schotman, P., and Gispen, W. H.,** Purification and some characteristics of an ACTH-sensitive protein kinase and its substrate protein in rat brain membranes, *J. Neurochem.*, 34, 1699, 1980.
50. **Aloyo, V. J., Zwiers, H., and Gispen, W. H.,** Phosphorylation of B-50 protein by calcium-activated, phospholipid-dependent protein kinase and B-50 protein kinase, *J. Neurochem.*, 41, 649, 1983.
51. **Lynch, T. J., Lo, S. J., Taylor, J. D., and Tchen, T. T.,** Characterization of and normal effects on subcellular fractions from Zanthophores of the goldfish *Carassius auratus* L., *Biochem. Biophys. Res. Commun.*, 102, 127, 1981.
52. **Lynch, T. J., Taylor, J. D., and Tchen, T. T.,** Regulation of pigment organelle translocation. I. Phosphorylation of the organelle-associated protein p57, *J. Biol. Chem.*, 261, 4204, 1986.
53. **Lynch, T. J., Wu, B., Taylor, J. D., and Tchen, T. T.,** Regulation of pigment organelle translocation. II. Participation of a cAMP-dependent protein kinase, *J. Biol. Chem.*, 261, 4212, 1986.
54. **Jahn, R., Padel, U., Porsch, P. H., and Söling, H. D.,** Adrenocorticotropic hormone and α-melanocyte-stimulation hormone induce secretion and protein phosphorylation in the rate lacrimal gland by activation of a cAMP-dependent pathway, *Eur. J. Biochem.*, 126, 623, 1982.
55. **Jahn, R. and Söling, H. D.,** Protein phosphorylation during secretion in the rat lacrimal gland, *FEBS Lett.*, 131, 28, 1981.
56. **Jahn, R. and Söling, H. D.,** Phosphorylation of the same specific protein during amylase release evoked by adrenergic or cholinergic agonists in rat and mouse parotide glands, *Proc. Natl. Acad. Sci. U.S.A.*, 78, 6903, 1981.
57. **Mains, R. E., Eipper, B. A., and Ling, N.,** Common precursor to corticotropins, endorphins, *Proc. Natl. Acad. Sci. U.S.A.*, 74, 3014, 1977.
58. **Peng Loh, Y.,** Immunological evidence for two common precursors to corticotropins, endorphins and melanotropin in neurointermediate lobe of the toad pituitary, *Proc. Natl. Acad. Sci. U.S.A.*, 76, 796, 1979.
59. **Watson, S. J. and Akil, H.,** On the multiplicity of active substances in single neurons: β-endorphin and α-melanocyte-stimulating hormone as a model system, in *Hormones and the Brain*, M.T.P. Press, Lancaster, England, 1980, 73.
60. **Wiegant, V. M., Verhaagen, J., Aloyo, V., and Gispen, W. H.,** ACTH and signal transduction in the neuronal membrane, in *Central Actions of ACTH and Related Peptides, Symposia in Neuroscience*, Vol. 4, De Wied, D. and Ferrari, W., Eds., Springer-Verlag, Heidelberg, 1986, 79.
61. **De Graan, P. N. E., Oestreicher, A. B., Schrama, L. H., and Gispen, W. H.,** Phosphoprotein B-50: localization and function, *Prog. Brain Res.*, 69, 37, 1986.
62. **Grahame-Smith, D. G., Butcher, R. W., Ney, R. L., and Sutherland, E. W.,** Adenosine 3′,5′-cyclic monophosphate as the intracellular mediator of the action of adrenocorticotropic hormone on the adrenal cortex, *J. Biol. Chem.*, 242, 5535, 1967.
63. **Sala, G. B., Hayashi, K., Catt, K. J., and Dufau, M. L.,** Adrenocorticotropin action in isolated adrenal cells. The intermediate role of cyclic AMP in stimulation of corticosterone synthesis, *J. Biol. Chem.*, 254, 3861, 1979.
64. **Kojima, I., Kojima, K., and Rasmussen, H.,** Role of calcium and cAMP in the action of adrenocorticotropin on aldosterone secretion, *J. Biol. Chem.*, 260, 4248, 1985.

Chapter 4

MELANOTROPIN MECHANISMS OF ACTION: MELANOGENESIS

John Pawelek, John McLane, and Michael Osber

TABLE OF CONTENTS

I. Introduction 48

II. Isolation of Genetic Variants of the Cloudman Melanoma Line 48

III. The Role of cAMP in the Responses to MSH 49
- A. cAMP-Dependent PKs 49
- B. Role of Phosphoproteins 49

IV. New Regulators of the Melanogenesis Pathway 50
- A. Cytotoxicity of Melanin Precursors 50

V. Synthesis of ^{125}I-Labeled β-MSH for Analyses of MSH Receptors 51
- A. Cooperative Interactions Between Receptors for MSH 52
- B. Dopa Phosphates: A New Class of Compound which Regulates Pigment Biosynthesis and Cellular Responsiveness to MSH 53

VI. Conclusions 57

Acknowledgment 57

References 57

I. INTRODUCTION

In the early part of this century, it was discovered that experimental removal of pituitary glands from tadpoles resulted in rapid lightening of their integument and that addition of pituitary extracts to the water in which these animals were living reversed the effect, causing integumental darkening.[1,2] Subsequent studies revealed that the active ingredients of the pituitary extracts belonged to a class of closely related peptides today known interchangeably as melanotropins or melanocyte stimulating hormones or simply MSH. Melanotropins are synthesized as part of a large precursor molecule, proopiomelanocortin.[3] Three forms have been described, α, β, and γ, ranging in size from 12 to 18 amino acids. Each of the three peptides has sequence homology to adrenocorticotropic hormone (ACTH) and to β-lipotropin (β-LPH). MSH, as these peptides will henceforth be referred to, is active throughout the vertebrate kingdom in stimulating the pigmentary system[4,5] although the mechanisms of action differ between the lower vertebrates and mammals. In addition to stimulation of the pigmentary system, there are a number of other functions that might be assigned to MSH, e.g., in regulation of synaptic transmission[6] and in aldosterone biosynthesis in adrenal glomerulosa cells.[7] Whether or not a cell is able to respond to MSH is likely to depend on the ability to express high-affinity receptors for MSH and to exhibit biochemical pathways that are regulated by the formation of the MSH-receptor complexes.[8] In 1973, Wong and Pawelek[9] first demonstrated that Cloudman S91 mouse melanoma cells growing in culture respond to MSH with increases in pigmentation as well as alterations in morphology and rates of proliferation. Since that initial observation there have been a number of studies with these cells on various aspects of the mechanisms of action of MSH. Evidence has been obtained to support the following conclusions:

1. MSH binds to specific, high-affinity receptors on the cell surface.[10,11]
2. Formation of the hormone-receptor complex is followed by a stimulation of the adenylate cyclase system and a net increase in intracellular levels of (cAMP).[12,13]
3. The hormone-receptor complex is internalized and apparently migrates to premelanosomes, the intracellular sites of melanin synthesis.[8]
4. The increased levels of cAMP result in increased tyrosinase (EC 1.14.18.1) activity and melanin deposition as well as changes in morphology and rates of proliferation.[9]
5. These processes are mediated, at least in part, by cAMP-dependent protein kinases.[14-16]

This chapter will review evidence in support of the above conclusions. In addition, improved techniques for synthesis and storage of mono-^{125}I-β-MSH will be presented since this ligand has proven invaluable in studies on the structure and function of receptors for MSH.

II. ISOLATION OF GENETIC VARIANTS OF THE CLOUDMAN MELANOMA LINE

In 1973, it was first shown that MSH can reduce the growth rate of Cloudman melanoma cells in culture.[9] This observation opened the way for isolating mutant cell lines which were resistant to these suppressive effects of MSH on proliferation. Subsequent studies resulted in the establishment of techniques for isolating pure lines of cells that did not respond to MSH, and showed no changes in either pigment formation, morphology, or rates of proliferation.[17,18] Analyses of the defects in these mutants allowed us to conclude that there are common biochemical pathways for the control of different cellular functions by MSH. For example, cells selected for resistance to MSH in growth frequently also showed no increase in pigmentation in response to MSH even though no selective pressure had been applied for pigmentation. Since the appearance of a particular mutant trait is a rare event (one in 10^6) the chances of having a second trait occur as an independent mutation and be isolated accidentally with no selective pressure would be extremely rare, statistically $10^{-6} \times 10^{-6}$ or one in 10^{12}. It is, therefore, virtually impossible to isolate double mutants without selective pressure. The more likely conclusion is that a single mutational event affected both proliferation and pigmentation, and, reciprocally in "normal" melanoma cells, a common biochemical step controls both functions. One common biochemical pathway controlling response to MSH is the cAMP system.

III. THE ROLE OF cAMP IN THE RESPONSES TO MSH

The responses of MSH are apparently mediated through increases in intracellular levels of cAMP. MSH causes a rapid and dramatic increase in cAMP levels. The action of MSH is mimicked by analogs of cAMP, dibutyryl cAMP or 8-bromo cAMP, as well as agents that raise intracellular levels of cAMP, as cholera toxin, prostaglandin E_1, theophylline, or isobutylmethylxanthine. Some mutant cell lines that do not respond to MSH with changes in growth or pigmentation also do not show increases in intracellular levels of cAMP.[18]

A. cAMP-Dependent PKs

The observations of Kuo and Greengaard[79] on the widespread occurrence of cAMP-dependent PKs throughout the phyla, suggested that these enzymes are the mediators of the many and varied response to cAMP. This situation appears to be true, at least in part, for MSH and Cloudman melanoma cells. Two lines of evidence are as follows. First, it was shown that the activity of tyrosinase, a rate-limiting enzyme in melanin biosynthesis, is increased in intact cells by MSH or by analogs of cAMP.[9,18] In 1977, it was reported that this increased tyrosinase activity could be achieved in a test tube using extracts of melanoma cells, cAMP, ATP, magnesium, and a partially purified cAMP-dependent PK prepared from the same cells.[14] Second, it was shown that a cAMP-dependent PK from Cloudman cells is involved in the regulation of proliferation by MSH.[15] The involvement of cAMP-dependent PKs in these functions implies that protein phosphorylation/dephosphorylation reactions are part of the biochemical pathway influenced by MSH.[16]

B. Role of Phosphoproteins

A number of observations has led us to conclude that phosphoproteins are important in the regulation of proliferation and pigmentation. Indirect evidence, discussed above, comes from the findings that the action of MSH is mediated through cAMP-dependent PK activity. Through direct analyses, we have observed as many as eight proteins whose phosphorylation is regulated by MSH.[16] This phenomenon is highly complex in that there are several subsets of regulatory steps involved. Some proteins show decreased phosphorylation in lysates of cells exposed to MSH; others show increased phosphorylation. One group of proteins of approximately 47,000 daltons shows increased phosphorylation when cAMP is included in the in vitro reaction mix, but only in cells not previously exposed to MSH. None of the other MSH-regulated proteins is affected by cAMP in the reaction mix. Phosphorylation of three proteins is regulated "qualitatively" by MSH in that exposure to the hormone seems to result in a switch between phosphorylation of tyrosine residues and phosphorylation of serine and threonine residues. This finding could be extremely important because phosphorylation of tyrosine residues has recently been linked to the control of growth. The possibility is raised that the "phosphotyrosine proteins" are those involved in growth regulation while the others are involved in regulation of melanogenesis.[16]

IV. NEW REGULATORS OF THE MELANOGENESIS PATHWAY

Melanin is formed spontaneously from dopaquinone in a test tube. This fact led to a generally accepted conclusion that the only catalyzed steps in melanogenesis are the conversion of tyrosine to dopa and of dopa to dopaquinone by tyrosinase (EC 1.14.18.1, also referred to as "dopa oxidase"). However, recent studies, beginning with those of Logan and Weatherhead[20] on Siberian hamsters, have indicated that there are additional regulatory points governing the Mason-Raper pathway.[20-27] Using amelanotic Cloudman cells which lacked tyrosinase activity, we showed that dopachrome conversion factor accelerates the conversion of dopachrome into 5,6-dihydroxyindole-2-carboxylic acid and that indole-blocking factor inhibits the conversion of 5,6-dihydroxyindole (5,6-DHI) into melanochrome.[21-23] Also, we showed that melanotic cells possess an activity that accelerates the conversion of 5,6-DHI into melanochrome and that this activity can at least, in part, be accounted for by tyrosinase itself.[24] Therefore, this unusual enzyme can catalyze three reactions in the synthesis of melanin.[24] In addition we have observed in Cloudman cells and in skins from neonatal mice, an activity, distinct from tyrosinase, which also accelerates conversion of 5,6-DHI into melanochrome.[25] The nature of this activity is as yet unknown.

Barber et al.[26] have confirmed and extended our observations of dopachrome conversion factor and indole blocking factor activities. They have detected these activities in a variety of different sources, including human hair bulbs. Their studies have led them to conclude that the activities can be accounted for by a single enzyme. We have as yet been unable to separate dopachrome conversion factor and indole-blocking factor activities from each other, consistent with the "single enzyme" proposal of Barber et al. However, it should be pointed out that until the activities are purified to homogeneity, no definitive statement regarding this situation is possible.[27]

Taken together, the results from three separate laboratories support the notion that there are regulatory steps in the Mason-Raper pathway following the formation of dopaquinone. To our knowledge, these regulated steps have been observed exclusively in eumelanogenic systems, and it remains to be seen whether they also are involved in pheomelanogenesis. It is our feeling that many additional regulatory events in these pathways exist. In this regard, we found that in some, but not all experiments, melanin formed by extracts of skins from brown mice was brown, whereas melanin formed by extracts of skins from sombre mice and from black mice was always black.[25] This was surprising since differences in color between brown and black mice were considered due to size and shapes of melanosomes rather than color of the melanin.[28] Since we observed differences in color between the melanins synthesized in the absence of melanosomes, perhaps there are determinants in skin extracts which regulate formation of brown or black melanin.

A. Cytotoxicity of Melanin Precursors

From a variety of clinical and experimental observations, Lerner[29] proposed several years ago that during the process of melanization, a pigment cell produces substances which are potentially autotoxic. Since that proposal, several laboratories have confirmed that the precursors of melanin biosynthesis are highly toxic to melanoma cells, probably because of their ability to generate free radicals as they co-polymerize. We showed that tyrosine and dopa are toxic to melanoma cells in culture and that the toxicity is increased significantly if the cells are preexposed to MSH.[13,23] We also showed that other intermediates, i.e., dopachrome and 5,6-DHI, are even more toxic than tyrosine and dopa.[23,30] However, cells that synthesize melanin possess mechanisms for protecting themselves from the cytotoxic intermediates, and our evidence indicates that in Cloudman melanoma cells, the protective mechanism is related to the relative activities of tyrosinase, dopachrome conversion factor, and indole-blocking factor.

V. SYNTHESIS OF ^{125}I-LABELED β-MSH FOR ANALYSES OF MSH RECEPTORS

Freychet et al.[31] reported the synthesis of ^{125}I-labeled insulin in 1971. This achievement opened the way for studies on receptors for insulin and led to the development of techniques for iodination and isolation of many biologically active peptide hormones. Varga et al.[10] first reported the synthesis of ^{125}I-labeled β-MSH and the use of this ligand to investigate MSH receptors in cultured melanoma cells. Lambert and Lerner,[11] Lambert et al.,[32] and Lambert and Varga[33] later described an improved synthesis and purification of ^{125}I-β-MSH which resulted in the isolation of a radioactive ligand that was monoiodinated at a single tyrosine residue and that retained full biological activity when compared to nonradioactive β-MSH. Their methods employed the use of IodoGen® (1,3,4,6-tetrachloro-3,6-diphenylglycoluril, Pierce Chemicals, Inc.) as a catalyst during the iodination reaction and isolation of the ^{125}I-β-MSH by reversed phase high-performance liquid chromatography (HPLC). Recently, the number of studies on receptors for MSH has increased markedly as has a concomitant need for a supply of ^{125}I-MSH. Improvements on the original techniques have now led to methodology wherein millicurie amounts of ^{125}I-MSH can be isolated and stored for at least 6 weeks. The following outlines our current procedures for synthesis and storage of this important ligand.

The synthesis and purification of ^{125}I-β-MSH was adapted from the methods of Lambert and Lerner,[11] and portions of their procedure are paraphrased or reproduced verbatim in this section. IodoGen® (Pierce) and fine glass beads (3M Corp. Superbrite type 120-5005) were placed into a 12 × 75-mm borosilicate glass tube, and IodoGen® dissolved by the addition of 100 μℓ dichloromethane. Dichloromethane was then evaporated in a stream of Argon gas leaving IodoGen® plated onto the glass beads. The reaction was carried out on ice with stirring. The reaction mixture consisted of 5 mCi sodium (^{125}I) iodide 2.3 nmol; Amersham, IMS300) in 10 μℓ and 1 mg HPLC-purified β-MSH in 100 μℓ borate buffer (9.5 g boric acid and 9.5g sodium borate in 1 ℓ distilled water, pH 8.2). Iodination was terminated after 60 sec by removal of the reaction mixture from the 12 × 75 mm tube and adding it to a vial containing 80 mg dithiothreitol (DTT®, Boehringer, Mannheim) and 2 μℓ of 50 mg/mℓ gentamicin solution (Sigma). The tube with the IodoGen®-coated beads was washed twice with 150-μℓ aliquots of borate buffered saline (6.25 g boric acid, 9.6 g sodium borate, and 9.5g NaCl, pH 8.8) and washings added to the vial containing the DTT®. This solution was then passed through an 0.45-μm ACRO® LC3A filter (Gelman) into a sterile 15-mℓ plastic tube (Corning) and incubated 16 to 20 hr with DTT® at 37°F to reduce any oxidized methionine in the β-MSH. The solution was desalted by passing over a Sephadex® G-10 column (5-mℓ bed) packed in a 5-mℓ plastic serological pipet and eluted with distilled water. β-MSH and ^{125}I-β-MSH both eluted at the void volume, while unreacted Na^{125}I was retained on the column. The peak fractions were combined and injected onto an HPLC system (two Waters Assoc. M6000A pumps, 660 programmer, U6K injector, uBondpak C_{18} column [30 × 0.39 cm]). Mono-^{125}I-β-MSH eluted from the column using a linear gradient of acetonitrile (Baker, HPLC grade) 19 to 27% over 25 min, with 1% triethylammonium phosphate (TEAP) pH 3.0 (Pierce) as the aqueous solvent. The flow rate was 2.0 mℓ/min and 0.6-mℓ fractions were collected and tested for radioactivity with an LKB® Mini Gamma counter. Mono-^{125}I-β-MSH eluted between fractions 50 to 56 and unlabeled β-MSH eluted between fractions 29 to 35 (monitored at 214 nm) as shown in the paper by Lambert and Lerner.[11] Fractions containing ^{125}I-β-MSH were placed on a Sephadex® G-10 column (20-mℓ bed) packed in a 10-mℓ plastic serological pipet and eluted with 0.05 *M* potassium phosphate, pH 6.0, containing 10 m*M* DTT® in order to remove the TEAP/acetonitrile solvent. ^{125}I-β-MSH eluted at the void volume and the peak fractions were pooled. These fractions were collected in 12 × 75 mm glass tubes coated with Sigmacote® (Sigma). Pooled fractions were aliquoted

Table 1
THE EFFECT OF IODOGEN® ON THE YIELD OF ^{125}I-β-MSH

Method	% yield
Lambert and Lerner[11]	2.5 ± 0.6
1.0 mg IodoGen®/50 mg beads	10.5 ± 1.1
2.0 mg IodoGen®/100 mg beads	23.7 ± 0.1
4.0 mg Iodogen®/200 mg beads	14.8 ± 1.6

Note: Based on initial amount of ^{125}I, average of two syntheses.

(150 to 200 μℓ) into plastic microcentrifuge tubes and stored in one of three ways — refrigerated at 4°, frozen at −20°, or dried under a stream of nitrogen and stored at −20°. After 3 and 6 weeks, a portion of each aliquot was rechromatographed using the RP-HPLC method described above.

As seen in Table 1, use of IodoGen®-coated glass beads results in a tenfold increase in the yield of ^{125}I-β-MSH as compared to the method of Lambert and Lerner.[11] The optimal concentration of 2 mg IodoGen® with 100 mg glass beads was used in later routine synthesis of ^{125}I-β-MSH. This increased yield resulted in the need to make ^{125}I-β-MSH less often and thus decreased the exposure to radioactive iodine.

Figure 1 shows effects of various storage conditions after 3 weeks on stability of ^{125}I-β-MSH. Storage in refrigeration at 4° led to a considerable breakdown while storage as frozen aliquots or as dried aliquots stored at −20° led to very little breakdown. Figure 2 shows the effect of the various conditions after 6 weeks. Refrigeration storage led to more breakdown products at 6 weeks as compared to 3 weeks. Storage as frozen aliquots or as dried aliquots showed only slightly more breakdown at 6 weeks than at 3 weeks. Although storage as dried aliquots at −20° seemed somewhat better at preventing the breakdown of ^{125}I-β-MSH, the inconvenience of drying down the number of aliquots typically obtained (30 to 40) has led us to choose the method of storing aliquots in a frozen state.

A. Cooperative Interactions Between Receptors for MSH

In 1974, it was reported that at least some of the responses to MSH (e.g., increased cAMP production and tyrosinase activity) occur in the G2 phase of the cell cycle[34] and that the apparent reason for this cell-cycle restriction is that receptors for MSH are most active in the G2 phase.[10] However, in the earlier work, the mechanism(s) for increased receptor activity were not studied (i.e., increased number? increased affinity?). We recently examined these phenomena in detail. We found that by three separate methods of obtaining populations of cells in the G2 phase of their cycle — centrifugal elutriation or synchronization with either colchicine or thymidine — there was increased binding of MSH by cells in the G2- and possibly late S-phases of their cycle (Table 2). However, cultures of cells passing through their cycle in synchrony were quite different from nonsynchronized (random) cultures. Both synchronized and random cultures expressed receptors for MSH in the G2- and possibly late S-phases of their cycle, but synchronized cultures bound 15 to 20 times more MSH per cell than random cultures. This increased binding of MSH by synchronized cells was accompanied by an increase in tyrosinase activity, providing direct proof of a relationship between cellular binding of MSH and cellular responsiveness to MSH. Analyses by Scatchard and Hill methods indicated that receptors from synchronized cells exhibited positive cooperativity, while receptors from random populations of cells exhibited no cooperativity (see Figures 3 and 4).[35]

Understanding the nature of these differences in receptor populations should be of fun-

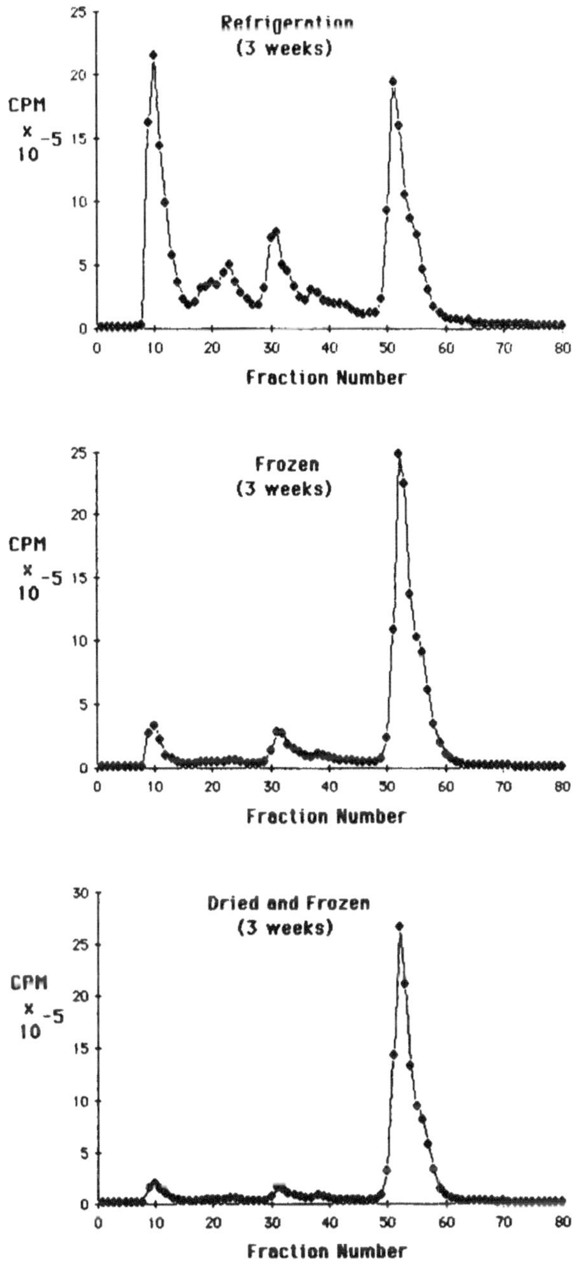

FIGURE 1. The HPLC profile of ^{125}I-β-MSH under various storage conditions after 3 weeks. Mono-^{125}I-β-MSH elutes between fractions 50 to 56.

damental value in understanding the mechanism of action of the MSH-receptor complex in regulating proliferation and pigmentation.

B. Dopa Phosphates: A New Class of Compound Which Regulates Pigment Biosynthesis and Cellular Responsiveness to MSH

One problem in many of the studies on the regulation of pigmentation is that a key metabolite, namely L-dopa, is relatively insoluble (maximum solubility is approximately 5 m*M* in aqueous solution at neutral pH) and is converted readily to melanin in the presence

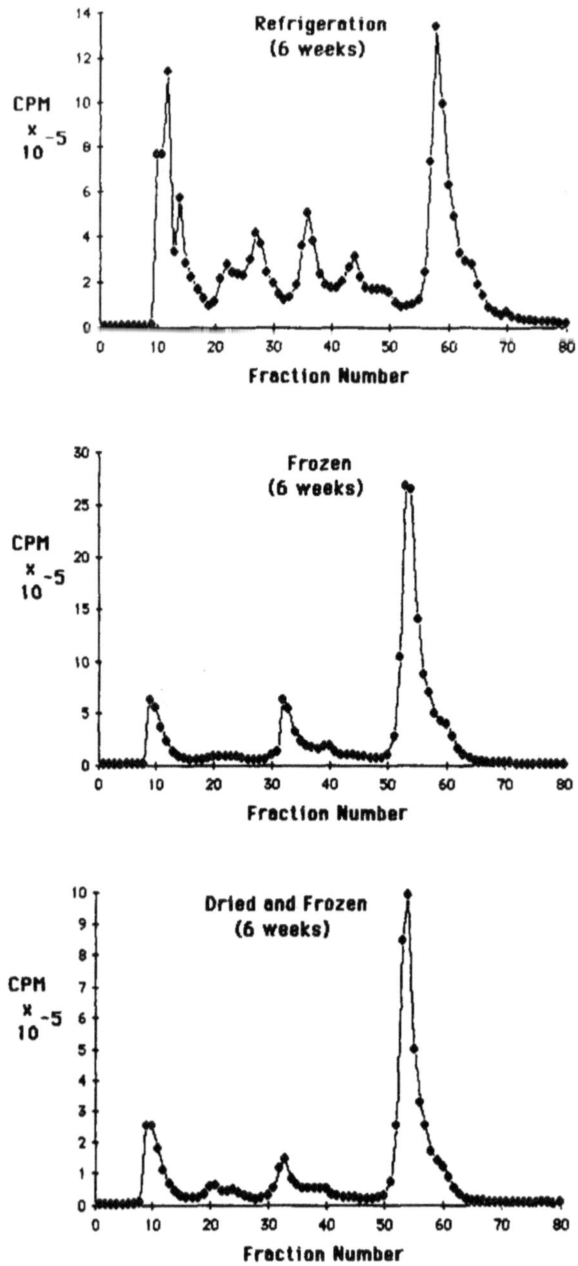

FIGURE 2. The HPLC profile of ^{125}I-β-MSH under various storage conditions after 6 weeks.

of O_2 above pH 7. Thus, many studies with cultured pigment cells or with melanomas in animals are difficult to carry out or to interpret when L-dopa is a component of the experiment. With this problem in mind, procedures were designed for attaching phosphate groups to the oxygens on the 3 and/or 4 positions of the phenylalanine ring.[36] The rationale was that such compounds would be stable to oxidation and far more soluble than dopa itself. It was reasoned that if melanocytes were able to take up these compounds, the cellular phosphatases might catalytically hydrolyze the phosphate groups and in the process produce free L-dopa. Results of our studies with these compounds are summarized as follows. Dopa phosphates are highly soluble compounds which are stable over a wide range of pH values and are not hydrolyzed

Table 2
BINDING OF ^{125}I-β-MSH TO CELLS SYNCHRONIZED WITH A DOUBLE THYMIDINE BLOCK

Hours after release from block	^{125}I-β-MSH bound (cpm/2.5 × 10^5 cells)
0	9,105 ± 2,068
6	22,288 ± 3,983
18	25,414 ± 4,291
24	33,648 ± 5,323
48	14,277 ± 1,114
70	5,313 ± 545

Note: Cells were synchronized wtih a double thymidine block as described by Puck.[40] At the times indicated, cells were collected and assayed for ^{125}I-β-MSH binding. Between 18 and 24 hr following release from the second thymidine block, more than 70% of the cells were in S and G2 ± M phases of the cycle as determined by analyses with a fluorescence-activated cell sorter. The results represent averages of three to five determinations from two experiments. The experiments were repeated several times with similar results.

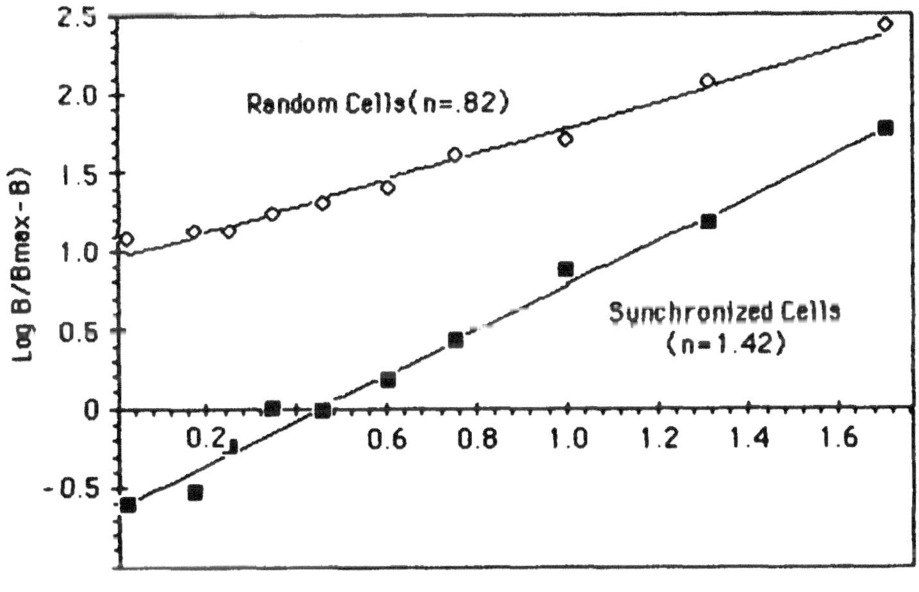

FIGURE 3. Specific binding of β-MSH to synchronized and random cells. The cells were synchronized by exposure to thymidine. The amount of MSH bound was plotted as a function of the amount of MSH added to the binding assay. Although the differences between the binding characteristics of synchronized and random cultures were qualitatively consistent between experiments, the absolute level of specific binding to the cells varied by as much as a factor of four from week to week. To compensate for this variability the data were normalized by adjusting four of the experiments on a percentage basis to the fifth experiment in which the absolute binding was highest. The normalized results represent the average and standard deviation for five separate experiments.

by boiling in concentrated acid. Synthetic yields of greater than 90% can be obtained using dopa as starting material. Exposure to alkaline phosphatase results in hydrolysis of the phosphate moieties and produciton of dopa. Dopa phosphates do not inhibit dopa oxidase (tyrosinase) activity. Dopa oxidase does not catalyze the conversion of dopa phosphates into melanin unless the dopa phosphates are first treated with phosphatase. Dopa phosphates,

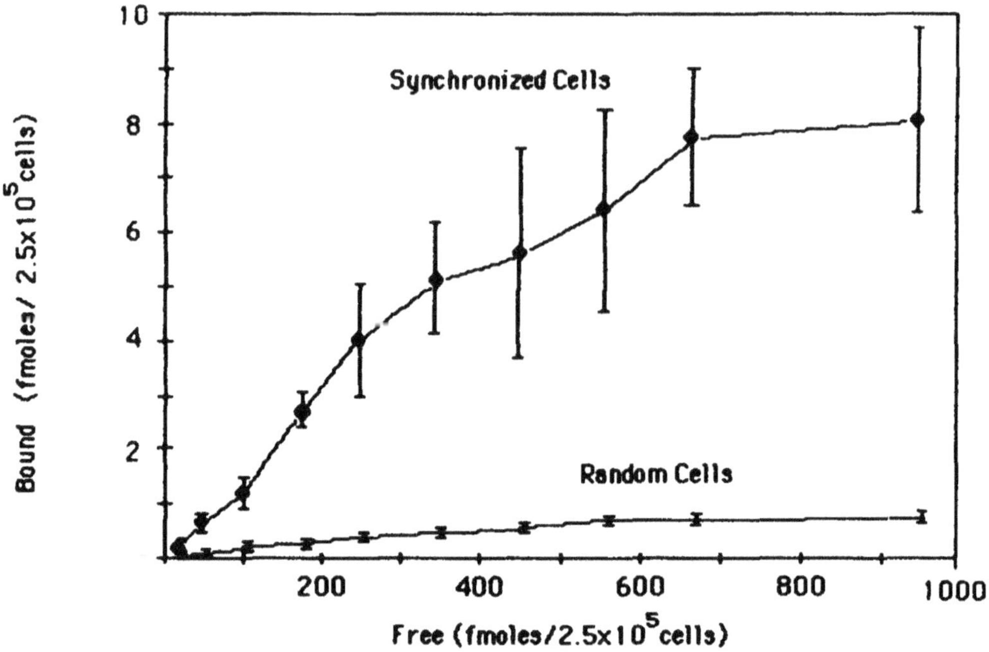

FIGURE 4. Hill analyses of the data in Figure 3. For synchronized cultures, the constant n = 1.42 indicates positive cooperativity between MSH receptors. For random cultures n = 0.82 indicates no cooperativity.

when compared to L-dopa, are stable in the presence of O_2 and are not oxidized by serum proteins. In the presence of cultured melanoma cells, dopa phosphates are readily converted into melanin, indicating that the cells are able to produce dopa from dopa phosphates. At high concentrations, dopa phosphates are cytotoxic toward melanoma cells in culture. The cytotoxicity is enhanced at least threefold by pretreatment of cells with melanotropin (MSH) and prevented by phenylthiourea as an inhibitor of dopa oxidase activity. These results, combined with studies on the uptake of radioactive forms of dopa phosphates (^{32}P and ^{14}C), indicate that phosphorylated isomers of dopa are efficiently taken up by Cloudman melanoma cells and are readily converted by the cells into a melanin precursor, presumably L-dopa.[37]

The above findings with dopa phosphates were obtained when the compounds were used at relatively high concentrations (10^{-3} *M*). A surprising finding emerged during investigations into possible effects of dopa phosphates on the activity of receptors for MSH.[37] Using ^{125}I-MSH as a tracer, we found that preincubation of Cloudman melanoma cells with very low concentrations of dopa phosphates (10^{-6} to 10^{-5} *M*) resulted in a three- to fourfold stimulation of MSH receptor activity. In addition, the stimulation of receptor activity was followed by a marked increase in MSH-mediated stimulation of tyrosinase activity and melanin deposition. Scatchard analyses indicated that the stimulation of receptor activity occurred through an increase in the number of the receptors for MSH rather than through an increase in the affinity of receptors. Apparently, at low concentrations, dopa phosphates act either directly or indirectly as cofactors for the MSH receptors. It is important to note that at the low concentrations dopa phosphates did not increase tyrosinase activity or melanin content, rather they increased the cellular responsiveness to MSH.[38]

The effects of dopa phosphates on melanoma cells in culture were consistent with findings in similar in vivo studies on SKH2 mice.[39] At high concentrations (0.1%) pigmentation was suppressed, while at low concentrations (0.01%) pigmentation was enhanced. It is possible that the reduction in skin pigment with high concentrations of dopa phosphates was due to the cytotoxic effects of melanin intermediates described above, while the stimulation of pigmentation with low concentrations of dopa phosphates was due to a stimulation of receptors for MSH.

VI. CONCLUSIONS

Hormonal regulation of cellular proliferation and the expression of differentiated functions in a number of cell types are topics of considerable attention. As with most biological studies, the rapid increase in the number of observations has underlined the diversity of the topics and raised more questions than orginally imagined. The studies on Cloudman melanoma cells and their regulation by MSH presented here represent a good example of this situation. Although we have made several observations on the regulation of proliferation and pigmentation, the findings have indicated that the regulation of these functions is highly complex and still poorly understood. Nonetheless, the fact that these cells grow in culture and respond to MSH, the ability to manipulate them genetically, the development of radioactive MSH as a tracer, and the development of new techniques such as production of monoclonal antibodies and the ability to clone genes, should all be of assistance in the future progress of work on this system.

ACKNOWLEDGMENT

Portions of this chapter were presented at the Aaron B. Lerner Symposium, Yale University in June 1985 and were published in the *Yale Journal of Biology and Medicine*, Volume 58, number 6, 1985.

REFERENCES

1. **Smith, P. E.,** Experimental ablation of the hypophysis in the frog embryo, *Science*, 44, 755, 1916.
2. **Atwell, W. J.,** On the nature of the pigmentation changes following hypophysectomy in the frog larva, *Science*, 49, 58, 1919.
3. **Nakanishi, S., Inoue, A., Kita, T., Nakamura, M., Chang, A. C. Y., Cohen, S. N., and Numa, S.,** Nucleotide sequence of cloned cDNA for bovine corticotropin-βlipotropin precursor, *Nature (London)*, 278, 423, 1979.
4. **Bagnara, J. and Hadley, M.,** *Chromatophores and Color Change*, Prentice-Hall, Englewood Cliffs, New Jersey, 1973.
5. **Lerner, A. B. and McGuire, J.,** Effect of alpha- and beta-melanocyte stimulating hormones on the skin color of man, *Nature (London)*, 189, 176, 1961
6. **Krivoy, W. A., Couch, J. R., and Stewart, J. M.,** Modulation of spinal synaptic transmission by beta-melanocyte stimulating hormone (MSH), *Psychoneuroendocrinology*, 10, 103, 1985.
7. **Yamakado, M., Franco-Saenz, R., and Mulrow, P. J.,** Sites of action of beta-melanocyte stimulating hormone in aldosterone biosynthesis in rat, *Proc. Soc. Exp. Biol. Med.*, 179, 318, 1985.
8. **Lerner, A. B., Moellmann, G., Varga, J., Halaban, R., and Pawelek, J.,** Action of MSH on pigment cells, in *Hormones and Cell Culture*, Cold Spring Harbor Conf. on Cell Proliferation, Vol. 6, 1979, 187.
9. **Wong, G. and Pawelek, J.,** Control of phenotypic expression of cultured melanoma cells to melanocyte-stimulating hormones, *Nature (London) New Biol.*, 241, 213, 1973.
10. **Varga, J. M., DiPasquale, A., Pawelek, J., McGuire, J., and Lerner, A.,** Regulation of melanocyte-stimulating hormone (MSH) action at the receptor level: discontinuous binding of MSH to synchronized mouse melanoma cells during the cell cycle, *Proc. Natl. Acad. Sci. U. S. A.*, 71, 1590, 1974.
11. **Lambert, D. T. and Lerner, A. B.,** Optimization of a melanotropin-receptor binding assay using reversed-phase high performance liquid chromatography, *J. Chromatogr.*, 266, 567, 1983.
12. **Bitensky, M. W., Demopoulos, H. B., and Russell, V.,** MSH-responsive adenyl cyclase in the Cloudman S91 melanoma, in *Pigmentation: Its Genesis and Biologic Control*, Riley, V., Ed., Appleton-Century-Crofts, New York, 1972, 247.
13. **Pawelek, J., Wong, G., Sansone, M., and Morowitz, J.,** Molecular controls in mammalian pigmentation, *Yale J. Biol. Med.*, 46, 430, 1973.

14. **Korner, A. and Pawelek, J.**, Activation of melanoma tyrosinase by a cyclic AMP-dependent protein kinase in a cell-free system, *Nature (London)*, 267, 444, 1977.
15. **Pawelek, J. M.**, Evidence suggesting that a cyclic AMP-dependent protein kinase is a positive regulator of proliferation in Cloudman S91 melanoma cells, *J. Cell. Physiol.*, 98, 619, 1979.
16. **Pawelek, J., Fleischmann, R., McLane, J., Guillette, B., Emanuel, J., Korner, A., Bergstrom, A., and Murray, M.**, Studies on growth and pigmentation of Cloudman S91 melanoma cells, *Pigment Cell*, 7, 521, 1985.
17. **Pawelek, J., Sansone, M., Morowitz, J., Moellmann, G., and Godawska, E.**, Genetic control of melanization: isolation and analysis of amelanotic variants from cultured melanotic melanoma cells, *Proc. Natl. Acad. Sci. U.S.A.*, 71, 1073, 1974.
18. **Pawelek, J., Sansone, M., Koch, N., Christie, G., Halaban, R., Hendee, J., Lerner, A., and Varga, J.**, Melanoma cells resistant to inhibition of growth by melanocyte-stimulating hormone, *Proc. Natl. Acad. Sci. U.S.A.*, 72, 951, 1975.
19. **Kuo, J. F. and Greengard, P.**, Cyclic nucleotide-dependent protein kinases. IV. Widespread occurrence of adenosine 3′,5′ monophosphate-dependent protein kinase in various tissues and phyla of the animal kingdom, *Proc. Natl. Acad. Sci. U.S.A.*, 64, 1349, 1969.
20. **Logan, A. and Weatherhead, B.**, Pelage color cycles and hair follicle tyrosinase activity in the Siberian hamster, *J. Invest. Dermatol.*, 71, 295, 1978.
21. **Korner, A. and Pawelek, J.**, Dopachrome conversion: a possible regulatory point in melanin biosynthesis, *J. Invest. Dermatol.*, 75, 192, 1980.
22. **Korner, A. M. and Gettins, P.**, Synthesis *in vitro* of 5,6-dihydroxyindole-2-carboxylic acid by dopachrome conversion factor from Cloudman S91 melanoma cells, *J. Invest. Dermatol.*, 85, 229, 1985.
23. **Pawelek, J., Korner, A., Bergstrom, A., and Bologna, J.**, New regulators of melanin biosynthesis and the autodestruction of melanoma cells, *Nature (London)*, 286, 617, 1980.
24. **Korner, A. and Pawelek, J.**, Mammalian tyrosinase catalyses three reactions in biosynthesis of melanin, *Science*, 217, 1163, 1982.
25. **Murray, M., Pawelek, J., and Lamoreux, M. L.**, New regulatory factors for melanogenesis: developmental changes in neonatal mice of various genotypes, *Dev. Biol.*, 100, 120, 1983.
26. **Barber, J. I., Towsend, D., Olds, D. P., and King, R. A.**, Dopachrome oxidoreductase: a new enzyme in the pigment pathway, *J. Invest. Dermatol.*, 83, 145, 1984.
27. **Pawelek, J. M.**, A new enzyme in the pigment pathway?, *J. Invest. Dermatol.*, 84, 234, 1985.
28. **Silvers, W. K.**, *The Coat Colors of Mice: A Model for Mammalian Gene Action and Interaction*, Springer-Verlag, Basel, 1979.
29. **Lerner, A. B.**, On the etiology of vitiligo and gray hair, *Am. J. Med.*, 51, 141, 1971.
30. **Pawelek, J. and Lerner, A. B.**, 5′ 6-Dihydroxyindole is a melanin precursor exhibiting potent cytotoxicity, *Nature (London)*, 276, 627, 1978.
31. **Freychet, P., Roth, J., and Neville, D.**, Insulin receptors in the liver: specific binding of ^{125}I-insulin to the plasma membrane and its relation to insulin bioactivity, *Proc. Natl. Acad. Sci. U.S.A.*, 68, 1833, 1971.
32. **Lambert, D. T., Stachelek, C., Varga, J. M., and Lerner, A. B.**, Iodination of β-melanotropin: time course analysis of reaction mixtures by high pressure liquid chromatography and characterization of biologically active mono- and diiodo-β-melanotropin, *J. Biol. Chem.*, 257, 8211, 1982.
33. **Lambert, D. T. and Varga, J. M.**, in *Pigment Cell, Proc. 11th Int. Pigment Cell Conf.*, Seiji, M., Eds., University of Tokyo Press, 1981, 347.
34. **Wong, G., Pawelek, J., Sansone, M., and Morowitz, J.**, Response of mouse melanoma cells to melanocyte-stimulating hormone. Localization in the G-2 phase of the cell cycle, *Nature (London)*, 248, 351, 1974.
35. **McLane, J. A. and Pawelek, J. M.**, Specific binding of ^{125}I-MSH to Cloudman melanoma cells during the G2 phase of the cell cycle, *Fed. Proc. Fed. Am. Soc. Exp. Biol.*, 44, 498a, 1985.
36. **Pawelek, J.**, Diphosphophenylalanine: a new compound exhibiting selective toxicity toward melanoma cells, *J. Invest. Dermatol.*, 78, 357, 1982.
37. **Pawelek, J. and Murray, M.**, Phosphorylated isomers of L-dopa increase melanin formation and promote cytotoxicity in cultured melanoma cells, *Cancer Res.*, 46, 493, 1986.
38. **McLane, J., Murray, M., Osber, M., and Pawelek, J.**, Phosphorylated isomers of L-dopa stimulate MSH binding activity and responsiveness to MSH in cultured melanoma cells, *J. Invest. Dermatol.*, 87, 421, 1986.
39. **Agin, P. P., Sayre, R. M., and Pawelek, J. M.**, Phosphorylated mixed isomers of L-dopa increase melanin content in skins of SKH-2 pigmented hairless mice, *Photochem. Photobiol.*, 41s, 73, 1985.
40. **Puck. T. T.**, Studies on the life cycle of mammalian cells, *Cold Spring Harbor Symp. Quant. Biol.*, 27, 167, 1964.

Chapter 5

MELANOTROPIN RECEPTORS AND SIGNAL TRANSDUCTION

Tomi K. Sawyer, Mac E. Hadley, and Victor J. Hruby

TABLE OF CONTENTS

I. Introduction 60
  A. Peptide Hormone Mechanisms of Action 60

II. MSH-Receptor-Mediated Melanotropic Activity 60
  A. Evidence for MSH Receptors on Melanocytes 61
  B. MSH Radioligands and Radioreceptor Assays 62
  C. MSH Stimulation of Adenylate Cyclase 63
  D. Calcium Requirement for MSH Bioactivity 64

III. Perspectives 65
  A. A Molecular Model of MSH Mechanism(s) of Action 67
  B. MSH-ACTH Structure-Activity and Mechanism(s) of Action Relationships 69
  C. Heterospecific Regulation of MSH Bioactivity 69

References 71

I. INTRODUCTION

A general mechanism of biological information transfer for physiologically active molecules having endocrine, neuroendocrine, or neurotransmitter function is schematically illustrated in Figure 1. In this simple two-step model, an extracellular ligand may elicit agonist bioactivity by *binding* to a specific receptor which is localized in the plasma membrane component of the target cell and *signal transduction* of the ligand-receptor complex via interactions with other membrane-bound proteins (enzymes) resulting in the modulation of intracellular second messenger concentrations. Although a plethora of scientific articles have been published in this field of research,[1-12] the study of receptors and ligand-receptor complexes remains naive in terms of experimental details and theoretical understanding. Nevertheless, the receptor concept has evolved considerably since the original studies of Langley,[13] Ehrlich,[14] and Clark[15] introduced the subject. Pharmacological characterization of receptors has generally focused on the objectives of demonstrating ligand specificity and quantitatively determining ligand affinities, saturability, and reversibility by the application of radiolabeled ligand derivatives. With few exceptions, these receptor studies are performed using intact tissue, dispersed or cultured cells, or partially purified plasma membranes isolated from fractionated cell sources. Correlation of binding to *triggering* of ligand-receptor-mediated signal transduction is then analyzed by measurement of potential receptor-coupled enzyme processes (e.g., cyclic adenosine monophosphate [cAMP] or cGMP production, phosphatidylinositol metabolism, Ca^{2+} influx or efflux, protein kinase [PK] phosphorylation), or cellular/tissue bioactivities (e.g., secretion, contraction-relaxation, or polarization-depolarization). However, the molecular mechanism(s) of the ligand-receptor binding and triggering remain ambiguous. The identification of ligand analogs having agonist, partial agonist, or antagonist biological properties has provided useful probes to investigate these phenomena. Furthermore, three-dimensional structure of the receptor itself and the structural make-up of the plasma membrane milieu require more experimental and theoretical study. With these facts in mind, this chapter will focus on α-melanocyte-stimulating hormone (α-MSH) and its known melanotropic activities within the framework of peptide hormone-receptor interactions.

A. Peptide Hormone Mechanisms of Action

Peptide hormones are physiologically active molecules important in a variety of endocrine, neuroregulatory, and other cellular processes. Structurally, linear and cyclic peptides are composed of an ordered, covalent sequence of amino acids which ultimately defines specific topological and, in a few well-studied examples, conformational features essential to their molecular action.[16] Functionally, the biological activity of many peptide hormones on their target cells has been shown to involve interaction with a plasma-membrane-localized, complex regulatory enzyme system including: (1) high-affinity, hormone-specific receptors; (2) guanyl nucleotide-binding proteins; (3) monovalent and divalent metal-binding proteins; and (4) adenylate cyclase. In addition, the phospholipid milieu of the plasma membrane may also provide a regulatory role in biological signal transduction by coordinating the activities of the functional components of this enzyme system. The molecular basis of such peptide-receptor binding and subsequent signal transduction processes are elaborated further in Table 1.

II. MSH-RECEPTOR-MEDIATED MELANOTROPIC ACTIVITY

In retrospect, the study of MSH-stimulated melanocyte activities (e.g., melanosome dispersion and melanin biosynthesis) has integrated the expertise of endocrinologists, cell biologists, biochemists, molecular biologists, and organic and pharmaceutical chemists.

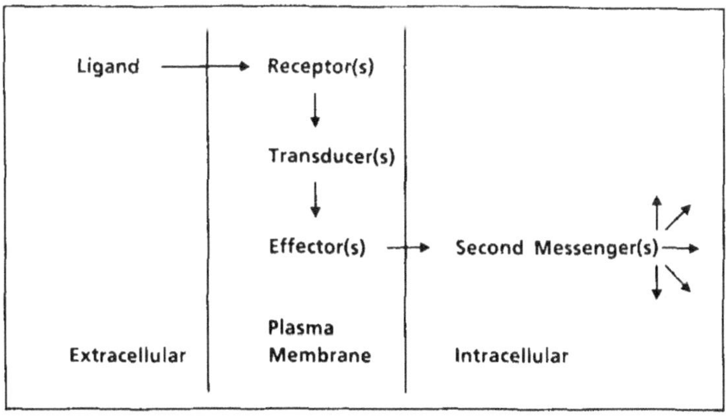

FIGURE 1. General scheme for biological information transfer at plasma membrane receptors by physiologically active ligands.

Table 1
KNOWN PROPERTIES OF PEPTIDE-RECEPTOR BINDING AND TYPES OF SIGNAL TRANSDUCTION

I. Properties of receptor binding
   A. Specific, capacity
   B. Affinity, on-off rates
   C. Molecular triggering (agonists vs. antagonist)
   D. Molecular interactive forces (electrostatic, van der Waals, hydrophobic, and hydrogen bonding)
   E. Lock-and-key vs. zipper theories
   F. Membrane (lipid biolayer) catalysis
   G. Homospecific, heterospecific, and allosteric regulation
   H. Internalization, turnover, degradation, and receptor biosynthesis
II. Types of receptor-mediated signal transduction
   A. Adenylate cyclase stimulation and cAMP biosynthesis
   B. Guanylate cyclase stimulation and cGMP biosynthesis
   C. Arachidonic acid metabolism
   D. Inositol phospholipid metabolism
   E. Calcium, ion "channeling" and/or calmodulin stimulation

Several in vitro and in vivo bioassays have been used (see Hadley and Castrucci, Chapter 1, this Volume), to investigate MSH action, and these include: (1) frog (*Rana pipiens*) skin bioassay; (2) lizard (*Anolis carolinensis*) skin bioassay; (3) tadpole (*Xenopus laevis*) skin bioassay; (4) fish (*Synbranchus marmoratus*) skin bioassay; and (5) murine melanoma (Cloudman S-91 or B-16) tyrosinase, protein kinase, or adenylate cyclase assays. The development of a reproducible and facile MSH radioreceptor assay has provided a significant challenge to be included in this biological testing strategy. However, radiolabeled (e.g., ^{3}H and ^{125}I) and biologically active derivatives of MSH have been recently reported, and available data are presented (*vide infra*) in this review.

A. Evidence for MSH Receptors on Melanocytes

It is generally believed that MSH interacts with plasma-membrane MSH-specific receptors rather than intracellular receptors which exist for steroid and thyroid hormones. Evidence for such MSH receptors was first indicated by studies on covalent MSH-macromolecular complexes (Table 2) which would unlikely be transported (i.e., internalized) across the plasma membrane. For example, derivatives of both α- and β-MSH have been chemically

Table 2
REPORTED EXAMPLES OF MSH-MACROMOLECULAR COMPLEXES AND THEIR BIOLOGICAL ACTIVITIES

I. MSH-polysaccharide conjugate
  A. β_p-MSH-Sepharose®
    Stimulated murine (Cloudman S-91) melanoma tyrosinase[17]
II. MSH-protein conjugates
  A. [D-Ala^1,Gly^2,Nva^3]α-MSH-human serum albumin
    Stimulated melanosome dispersion of frog (*R. pipiens*) skin melanocytes[18]
  B. α-MSH-ferritin-FITC
    Stimulated murine melanoma tyrosinase and melanosome dispersion on frog (*R. pipiens*) skin melanocytes[19]
III. MSH-virus conjugate
  A. [N-α-maleimidohexanoyl]α-MSH-TMV
    Stimulated prolonged melanosome dispersion on frog (*R. pipiens*) skin melanocytes and demonstrated high potency on murine (Cloudman S-91) melanoma melanocytes as well as on the frog skin bioassay[20]

conjugated to polysaccharides (i.e., sepharose[17]); proteins (i.e., albumin,[18] and ferritin[19]); and virus (i.e., tobacco mosaic virus (TMV).[20-22] Each of these MSH-macromolecular complexes was biologically active, and, interestingly, the MSH-TMV conjugates possessed higher potency and prolonged (residual) melanotropic activity compared to their MSH monomeric (unconjugated) precursors. The rapid reversal of melanotropic activity in vitro effected by naturally occurring α-MSH[20,23] also provides, perhaps, indirect evidence that cell-surface receptors exist, and suggests that MSH-receptor interactions are kinetically controlled by extracellular concentrations of α-MSH. The prolonged bioactivities of the MSH-TMV conjugates suggest that multiple ligand-receptor interactions may occur to modify the on-off (association-dissociation) rate of peptide hormone binding. Similar prolonged bioactivities have been reported[24] for [p-azido-Phe^{13}]-α-MSH-stimulated melanocytes (*X. laevis*) subsequent to photoaffinity labeling in which the MSH derivative was covalently bound (i.e., negligible ligand-receptor dissociation). Finally, iontophoretic injection of MSH into single melanocytes (*R. pipiens*) has been reported[25] to have no effect on melanosome dispersion.

The isolation and characterization of purified MSH receptors has not yet been documented, but such research is in progress. (Schwyzer and co-workers, personal communication). This may provide insight to understanding the molecular basis of MSH specificity at melanocytes vs. other potential MSH target systems (i.e., adrenal gland, adipose tissue, and the brain) which are also adrenocorticotropin (ACTH) sensitive. Recently, the adrenal cell ACTH receptor has been purified, and the molecular structure and binding characteristics determined.[26] A relevant question remaining to be answered is whether or not the structural homology between MSH and ACTH may also exist between MSH receptors and ACTH receptors. It is known that certain functional similarities exist between these two receptor types (*vide infra*) although the structure-activity relationships for α-MSH are quite different from those described for ACTH.[27]

B. MSH Radioligands and Radioreceptors Assays

Several radiolabeled MSH analogs have been prepared over the past decade to investigate MSH receptors on both murine and human melanoma cells in vitro and in vivo (Table 3). To date, however, there exists a paucity of biological data for these MSH radioligands. However, based on the above studies using ^{3}H-labeled[28,29] and ^{125}I-labeled[17,30-33] MSH analogs, it has been reported that approximately 10^4 MSH receptors per cell exist on murine melanoma cells. Future structure-activity studies utilizing such MSH radioreceptor assays are expected to have several applications in the examination of the molecular basis of MSH

Table 3
REPORTED EXAMPLES OF RADIOLABELED MSH DERIVATIVES AND THEIR BIOLOGICAL ACTIONS

I. [$^3H_2$]-MSH Analogs

   A. [($^3H_2$)Tyr^2]α-MSH; [($^3H_4$)NvA^{13}]α-MSH; and [(^{125}I)Tyr^2,($^3H_4$)Nva^{13}]α-MSH^{13}]α-MSH

      Demonstrated "specific" binding (10^{-10} to 10^{-8} *M* range) and estimated 10,000 to 30,000 receptors per cell on a nonsynchronized murine (Cloudman S-91) melanoma cell line[28]

   B. Ac-[(^{3}H)Nle^4, D-Phe^7]α-MSH$_{4-11}$-NH$_2$

      An enzymatically stable (vs. chymotrypsin, trypsin, and rat serum) and superpotent α-MSH congener which effects prolonged bioactivities in vitro (*R. pipiens* and *A. carolinensis* melanophores and murine [Cloudman S-91] melanoma cells) and in vivo (*X. laevis*) and may be useful for MSH receptor localization (i.e., diagnosis of melanoma tumors in vivo)[29]

II. (^{125}I)-MSH Analogs

   A. [(^{125}I)Tyr^5]β-MSH and [($^{125}I_2$)Tyr^5]β-MSH

      Demonstrated "specific" binding with an affinity constant ≈ 3×10^8 ℓ/mol, and estimated 10,000 receptors per cells on a synchronized murine (Cloudman S-91) melanoma cell line[30]

   B. [(^{125}I)Tyr^2, Nle^4]α-MSH

      A chemically stable (vs. Met oxidation and subsequent biological inactivation of Met-substituted MSH derivatives) α-MSH congener which is currently being used to investigate MSH receptors for murine (Cloudman S-91) melanoma cells in vitro[32]

   C. [(^{125}I)Tyr^2]α-MSH

      Demonstrated "specific" binding (based on isotopic dilution with unlabeled α-MSH, β-MSH, γ-MSH, ACTH, and MSH-related fragment peptides) with an affinity constant ≈ 1×10^7 ℓ/mol (Scatchard analysis) on two nonsynchronized human melanoma cell lines[33]

binding and *signal transduction*. These include: (1) designing MSH competitive antagonists; (2) investigating the prolonged bioactivities of MSH-TMV conjugates, photoaffinity-labeled MSH derivatives, and [Nle^4,D-Phe^7]-substituted analogs (*vide infra*) of α-MSH; and (3) identifying the role(s) of calcium as well as other heterotropic modulators of α-MSH-stimulated melanotropic activities (e.g., adenylate cyclase, tyrosinase, and melanosome maturation and secretion) on mammalian melanocytes. The development of the enzymatically stable and highly potent radiolabeled MSH analog Ac-[(^{3}H)Nle^4,D-Phe^7]α-MSH$_{4-11}$-NH$_2$[29] and/or congeners thereof, offers potential application to studies in vivo to localize melanoma tumors.

C. MSH Stimulation of Adenylate Cyclase

Signal transduction of the MSH-receptor complex to adenylate cyclase has been supported by several experimental data:

1. α-MSH stimulates intracellular cAMP production in frog (*R. pipiens*) melanocytes in vitro.[34]
2. Iontophoretically injected cAMP into single frog (*R. pipiens*) melanocytes effects melanosome dispersion.[25]
3. α-MSH stimulates cAMP production in both murine[35,36] and human[37] melanoma cells and stimulates adenylate cyclase activity dose-dependently in murine melanoma cell-free systems.[23,38,39]
4. cAMP and N^6,O$^{2'}$-dibutyryl cAMP mimic (presumably, after directly entering the cell) MSH-stimulated melanosome disperison in vitro on both frog and lizard skin melanocytes.[40-42]
5. Methylxanthines, which are inhibitors of cAMP phosphodiesterase (an intracellular cyclic nucleotide-degrading enzyme) mimic both MSH-stimulated melanosome dispersion on lizard (*A. carolinensis*) melanocytes[43] and MSH-stimulated tyrosinase activity in murine (Cloudman S-91) melanoma cells[44].

Table 4
COMPARATIVE PROPERTIES OF SEVERAL α-MSH ANALOGS ON THE LIZARD (*A. CAROLINENSIS*) SKIN BIOASSAY AND MURINE (CLOUDMAN S-91) MELANOMA ADENYLATE CYCLASE ASSAY

Compound	Potency relative to α-MSH	
	Lizard skin	Murine melanoma adenylate cyclase
1. α-MSH	1.0	1.0
2. Ac-[Nle^4]α-MSH$_{4-10}$-NH$_2$	0.06	0.09
3. Ac-[Nle^4,D-Phe^7]α-MSH$_{4-10}$-NH$_2$	7.5	7.7
4. Ac-[Nle^4]α-MSH$_{1-6}$-NH$_2$	Inactive	Inactive
5. Ac-α-MSH$_{11-13}$-NH$_2$	Inactive	Inactive
6. Ac-[Nle^4]α-MSH	1.5	2.0
7. Ac-[Nle^4,D-Phe^7]α-MSH	5.0	26.7
8. Ac-[Cys^4,Cys^{10}]α-MSH	30.0	3.0

Based on these data, a functional MSH receptor-transducer-effector system consisting of guanyl-nucleotide regulatory proteins (transducer) and adenylate cyclase (effector) is implicated as a mechanistic pathway for MSH-receptor-mediated signal transduction. Recently, it has been documented[45] that MSH structure-activity relationships on the lizard (*A. carolinensis*) skin bioassay correlate well with murine melanoma (Cloudman S-91) adenylate cyclase in vitro assays (Table 4) in terms of the relative potencies of several α-MSH fragments and [Cys^4,Cys^{10}] and [D-Phe^7]-substituted α-MSH analogs.

The functional role(s) of calcium (Ca^{2+}) in α-MSH-stimulated adenylate cyclase is also suggested based on several experimental data: (1) extracellular Ca^{2+} is specifically required for α-MSH, but not β-adrenergic agonists, prostaglandins, or cAMP derivatives, to stimulate melanosome dispersion in lizard (*A. carolinensis*) and frog (*R. berlandieri forrei*) skin melanophores in vitro[46,47] and (2) Ca^{2+} has been claimed[38] to be required for α-MSH stimulation of murine melanoma adenylate cyclase (Cloudman S-91); however, Ca^{2+} has also been reported[48] to be inhibitory in this melanoma adenylate cyclase assay. Other observations (*vide infra*) pertinent to Ca^{2+} and its functional role(s) in the mechanism(s) of action of MSH suggest that this hypothesis may be oversimplified.

D. Calcium Requirement for MSH Bioactivity

As previously stated, Ca^{2+} is specifically required for MSH-stimulated melanosome dispersion in several in vitro melanocyte systems (Table 5 and Figure 2). The Ca^{2+} ionophore, A23187, is also capable of mimicking this Ca^{2+} dependency for MSH-like melanotropic activity; however, studies[48] on the lizard (*A. carolinensis*) skin bioassay (Figure 3) revealed the unique properties (relative to α-MSH) of ionophore A23187 to elicit prolonged (residual) activity as well as its apparent ability to bind (or compartmentalize) to the melanocyte (presumedly, to the plasma membrane) in the absence of Ca^{2+}. Interestingly, these unique properties of ionophore A23187 also exist for [Nle^4,D-Phe^7]α-MSH (but again not for α-MSH) in terms of an apparent Ca^{2+} dependency for effecting prolonged activity of the lizard (*A. carolinensis*) skin melanophores, but Ca^{2+} independency for binding to MSH receptors[49] (Figure 4). The photoaffinity-labeled MSH derivative [Pap^{13}]-α-MSH has also been shown[50] to require Ca^{2+} for both MSH-receptor photolabeling and subsequent stimulation of tadpole (*X. laevis*) melanocytes. These data implicate functional roles of Ca^{2+} for both receptor *binding* and *signal transduction* of α-MSH. In contrast, the receptor-binding process of [Nle^4,D-Phe^7]α-MSH may be Ca^{2+} insensitive. Future studies on intact melanoma cells may facilitate a more thorough biochemical analysis of such MSH-Ca^{2+} relationships.

Table 5
SUMMARY OF MSH-SPECIFIC Ca^{2+} REQUIREMENTS FOR THE BIOLOGICAL ACTIONS OF MSH IN VITRO

	Tissue system	Bioactivity	Agonist Ca^{2+}	Requirement?
I.	Lizard (*A. carolinensis*), frog (*R. pipiens*), and toad (*Saphiopus couchi*) skin melanophores	Melanosome dispersion	α-MSH	Yes
			Prostaglandin E_2	No
			Isoproterenol	No
			Theophylline	No
II.	Lizard (*A. carolinensis*) skin melanophores	"Irreversible" melanosome dispersion	[Nle⁴, D-Phe⁷]α-MSH	Yes
			Ionophore A23187	Yes
III.	Tadpole (*X. laevis*) skin melanophores	Photoaffinity labeling of MSH receptors and "irreversible" melanosome dispersion	[*p*-Azido-Phe¹³]α-MSH	Yes
IV.	Tadpole (*X. laevis*) skin melanophores	Melanosome dispersion	α-MSH	Yes
			Forskolin	No

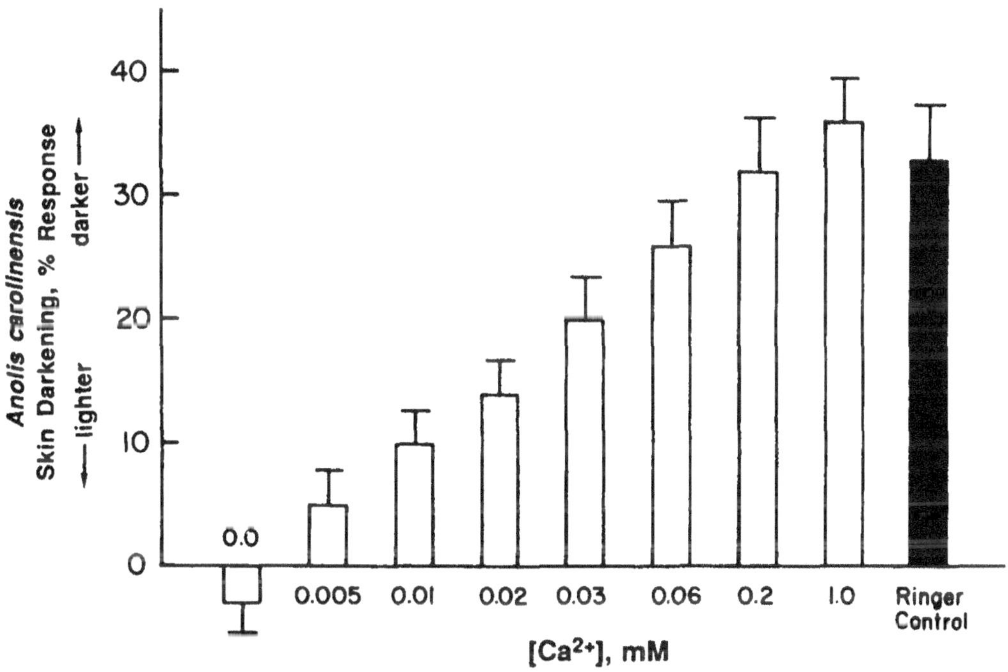

FIGURE 2. Dose-response effect of extracellular calcium (Ca^{2+}) on α MSH stimulated lizard (*A. carolinensis*) skin melanophores.

III. PERSPECTIVES

In retrospect, it should be noted that the subject of MSH receptors and biological mechanism(s) has been previously reviewed.[20,45,48,51-53] The objectives of this chapter have been to summarize recent advances in experimental findings which may delineate the molecular processes involved in MSH-receptor interactions important for both basic research and, possibly, therapeutic applications.

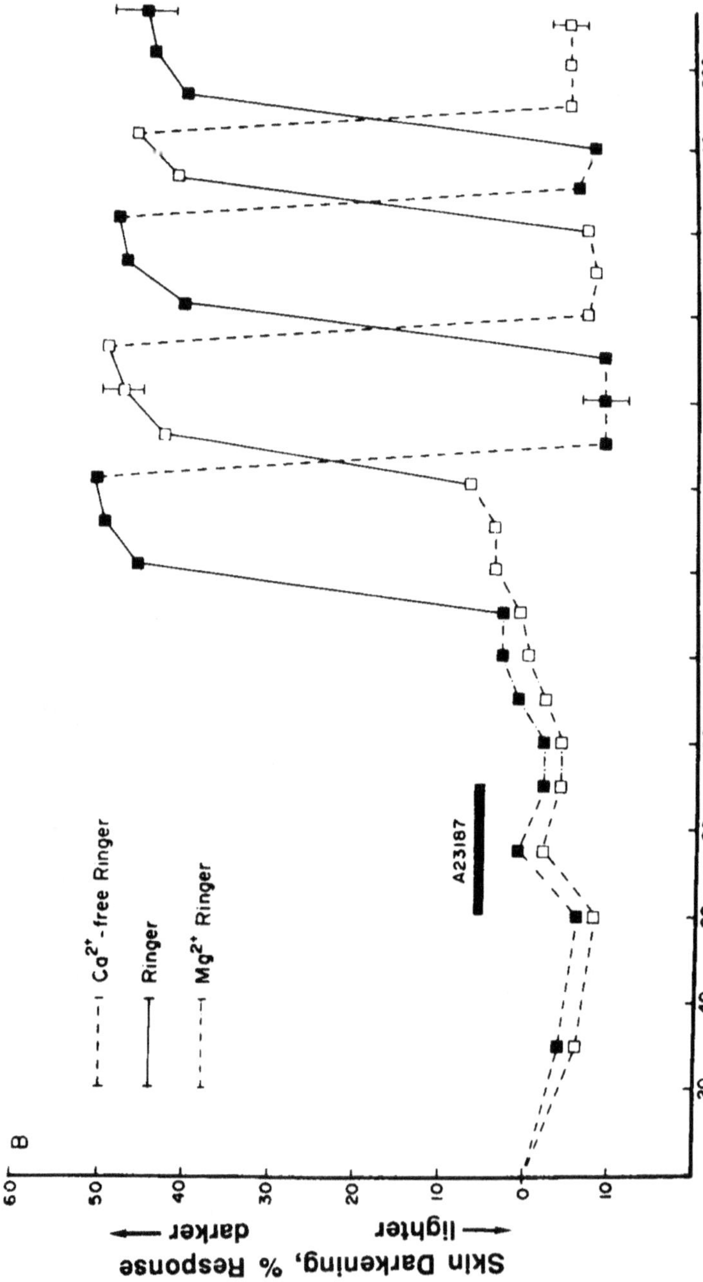

FIGURE 3. Effect of Ca^{2+} on ionophore A23187-stimulated lizard (*A. carolinensis*) skin melanophores

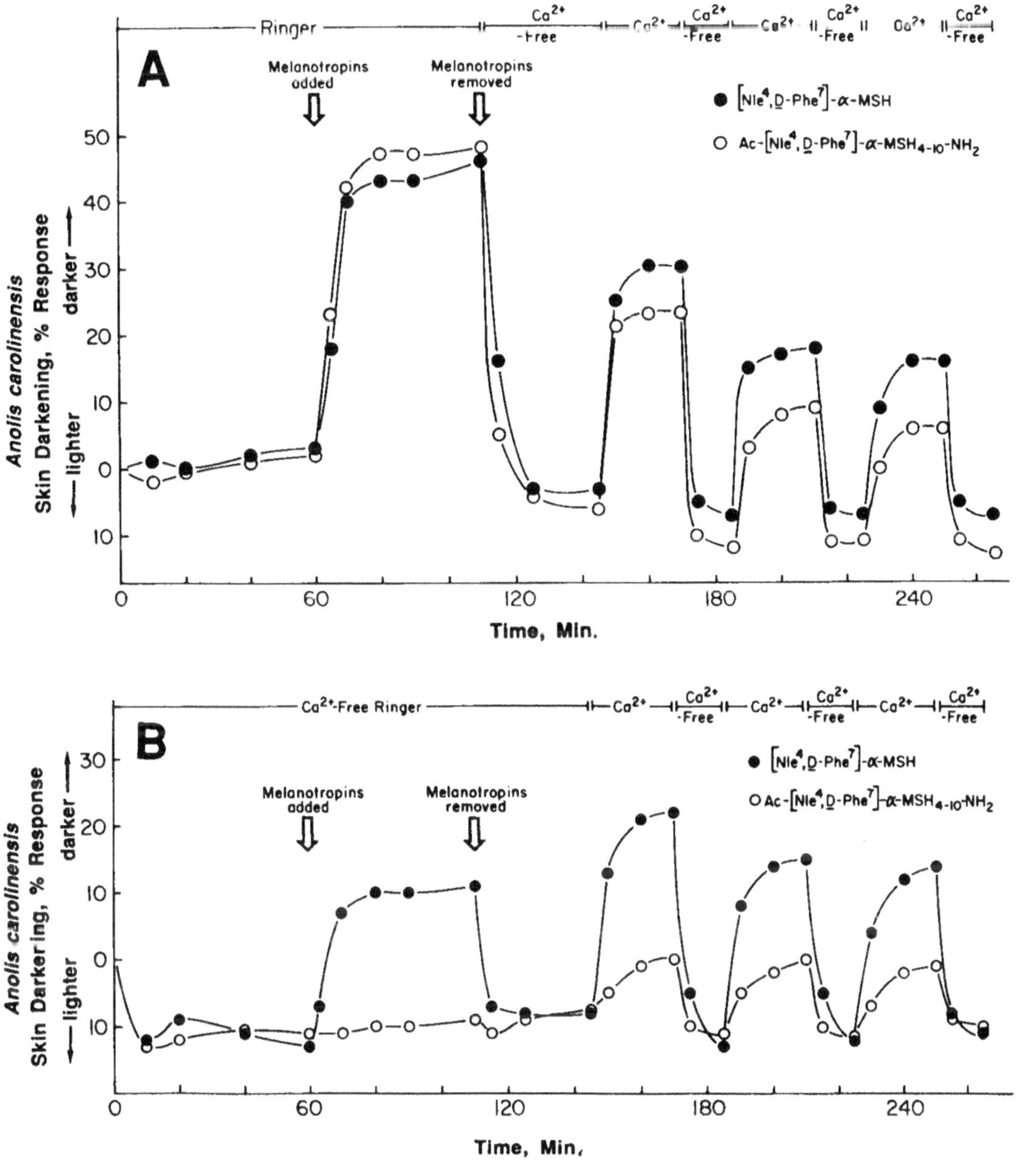

FIGURE 4. Effect of Ca^{2+} on [Nle^4,D-Phe^7]-α-MSH or Ac-[Nle^4,D-Phe^7]α-$MSH_{4\text{-}10}$-NH_2-stimulated lizard (*A. carolinensis*) skin melanophores.

A. A Molecular Model of MSH Mechanism(s) of Action

The development of MSH-based molecular probes (i.e., superpotent agonists, radiolabeled analogs, photoaffinity derivatives, and melanotropin-macromolecular conjugates) to evaluate the mechanism(s) of action of α-MSH has been an important hallmark in the progress of research on MSH receptors and signal transduction. However, current studies using such MSH-based molecular probes have not simplified our understanding of MSH-stimulated melanotropic activity. For example, it has been recently reported[54] that a MSH-TMV conjugate which effected superpotent agonism on a murine melanoma cell-line in stimulating tyrosinase activity was inactive in stimulating AMP production within these cells. This

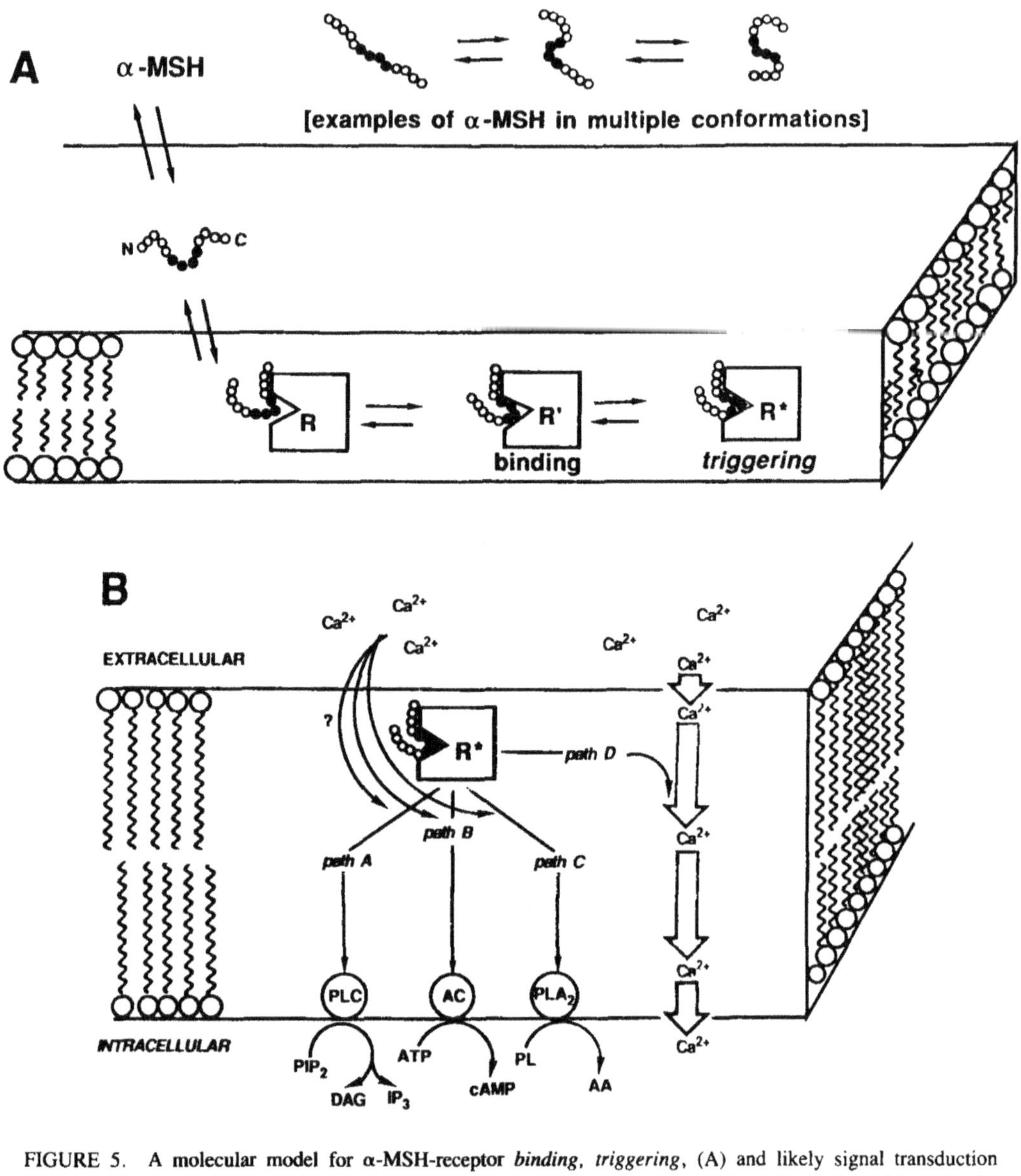

FIGURE 5. A molecular model for α-MSH-receptor *binding*, *triggering*, (A) and likely signal transduction pathways (B) to provide the melanotropic cellular responses (i.e., melanosome dispersion/melanin biosynthesis) effected by MSH peptides. Abbreviations are R, MSH receptor; AC, adenylate cyclase; ATP, 5'-adenosine triphosphate; cAMP, 3',5'-adenosine monosphosphate; PLC, phospholipase C; PIP_2, phophatidylinositol-4,5-bisphosphate; DAG, diacylglycerol; IP_3, inositol-1,4,5-triphosphate; PLA_2, phospholipase A_2; PL, phospholipid; and AA, arachidonic acid.

suggests that different MSH-receptor-mediated (i.e., other than adenylate cyclase) signal transduction pathways may be of physiological relevance to the native hormone. Likely candidates as such non-AMP second messengers may be diacylglycerol and/or Ca^{2+} as implicated by studies[55,56] with phorbal esters and inophore A23187-Ca^{2+} in the lizard (*A. carolinensis*) skin bioassay. A biological model for α-MSH *binding* and molecular *triggering* (signal transduction) of such melanocyte-membrane receptors is schematically illustrated in Figure 5. This model proposes that MSH-receptor binding might involve multiple, sequential steps including: (1) surface accumulation of the peptide by ionic interactions between its positively charged side-chain moieties (e.g., from the His-6, Arg-8, and Lys-11 residues of α-MSH) and the negatively charged phospholipids which constitute the membrane bilayer matrix; (2) secondary peptide-membrane interactions (i.e., hydrophobic and electrostatic) which provide regional, conformational, and orientational selectivity of the peptide prior to peptide-receptor recognition; and (3) specific peptide-receptor *binding* (i.e., α-MSH-melanotropin receptor complex formation). A similar biological model has recently been reported by Sargent and Schwyzer[57] for ACTH and dynorphin based on experimental studies of these amphiphilic peptides and synthetic lipid bilayers. Furthermore, we suggest that the MSH active site tetrapeptide sequence, His^6-Phe^7-Arg^8-Trp^9, may adapt a reverse-turn type secondary structure in its bioactive conformation at the melanocyte MSH receptor as previously described[58] for the cyclic, conformationally restricted α-MSH analog $[\overline{Cys^4,Cys^{10}}]$α-MSH (Figure 6).

B. MSH-ACTH Structure-Activity and Mechanism(s) of Action Relationships

In retrospect, the field of MSH research has diverged markedly from ACTH in terms of structure-conformation-activity relationships even though both peptide hormones share an identical N-terminal sequence, Ser^1-Tyr^2-Ser^3-Met^4-Glu^5-His^6-Phe^7]-Arg^8-Trp^9-Gly^{10}-Lys^{11}-Pro^{12}-Val^{13} (Figure 7). Exemplifying this divergence are [Nle^4,D-Phe^7][α-MSH[23] and $[\overline{Cys^4,Cys^{10}}]$α-MSH[58] which are quite potent MSH analogs but very weak ACTH congeners[59] on their respective target tissue systems. Interestingly, the success in developing[60] potent, low molecular weight, MSH-based derivatives (e.g., Ac-[Nle^4,D-Phe^7]α-$MSH_{4\text{-}9}$-NH_2; Ac-[D-Phe^7]α-$MSH_{5\text{-}9}NH_2$; Ac-[D-Phe^7]α-$MSH_{6\text{-}9}$-NH_2; and Ac-[D-Phe^7]α-$MSH_{7\text{-}9}$-NH_2) suggest that ACTH might yet be chemically modified to yield more simplified congeners. In contrast to the apparent divergence in MSH-ACTH peptides in structure-conformation-activity, similarities apparently exist for the two different peptide hormones in terms of receptor-mediated mechanism(s) of action. For example, it has recently been reported[58] that ACTH requires Ca^{2+} for receptor binding but not for adenylate cyclase stimulation. In addition, Ca^{2+} has been recently described[62] to be a synarchic messenger with cAMP via two hypothetically different ACTH-receptor types. These data are reminiscent of that observed (*vide supra*) for [Nle^4,D-Phe^7]α-MSH and ionophore A23187-Ca^{2+}. Again, there are many experimental details yet to be determined before a definitive model(s) for MSH-ACTH mechanism(s) of action can become well established.

C. Heterospecific Regulation of MSH Bioactivity

The heterospecific modulation of MSH-stimulated melanotropic activities has been extensively studied[63-66] using α-adrenergic agonists, dopamine, and melatonin as exemplified in Table 6. Although the mechanism(s) of action of these MSH inhibitors is presumably noncompetitive (i.e., not direct MSH-receptor-binding antagonism but, rather, an interruption of MSH-receptor-mediated signal transduction), there still exists a paucity of definitive experimental data to ascertain the anti-MSH effects of these compounds at the molecular level. Most recently, comparative studies[67-69] on melanin-concentrating hormone (MCH) on the fish (*S. marmortus*), lizard (*A. carolinensis*), and frog (*R. pipiens*) skin melanophore systems have demonstrated that this counter regulatory peptide hormone to α-MSH possesses profound structure-activity and mechanism of action properties which clearly are related to α-MSH, α-MSH · Ca^{2+}, and the α-melanotropin receptor.

In summary, it is our perspective that the study of MSH receptors remains a viable area for important discoveries in both basic research and therapeutic development. Such work is in progress at many scientific institutions, both academic and pharmaceutical, to further explore MSH or MSH-ACTH peptide effects on normal and transformed (cancerous) melanocytes as well as other known biological target tissue systems (see Volume II, and Chaturvedi and Hadley, Chapter 9, this Volume).

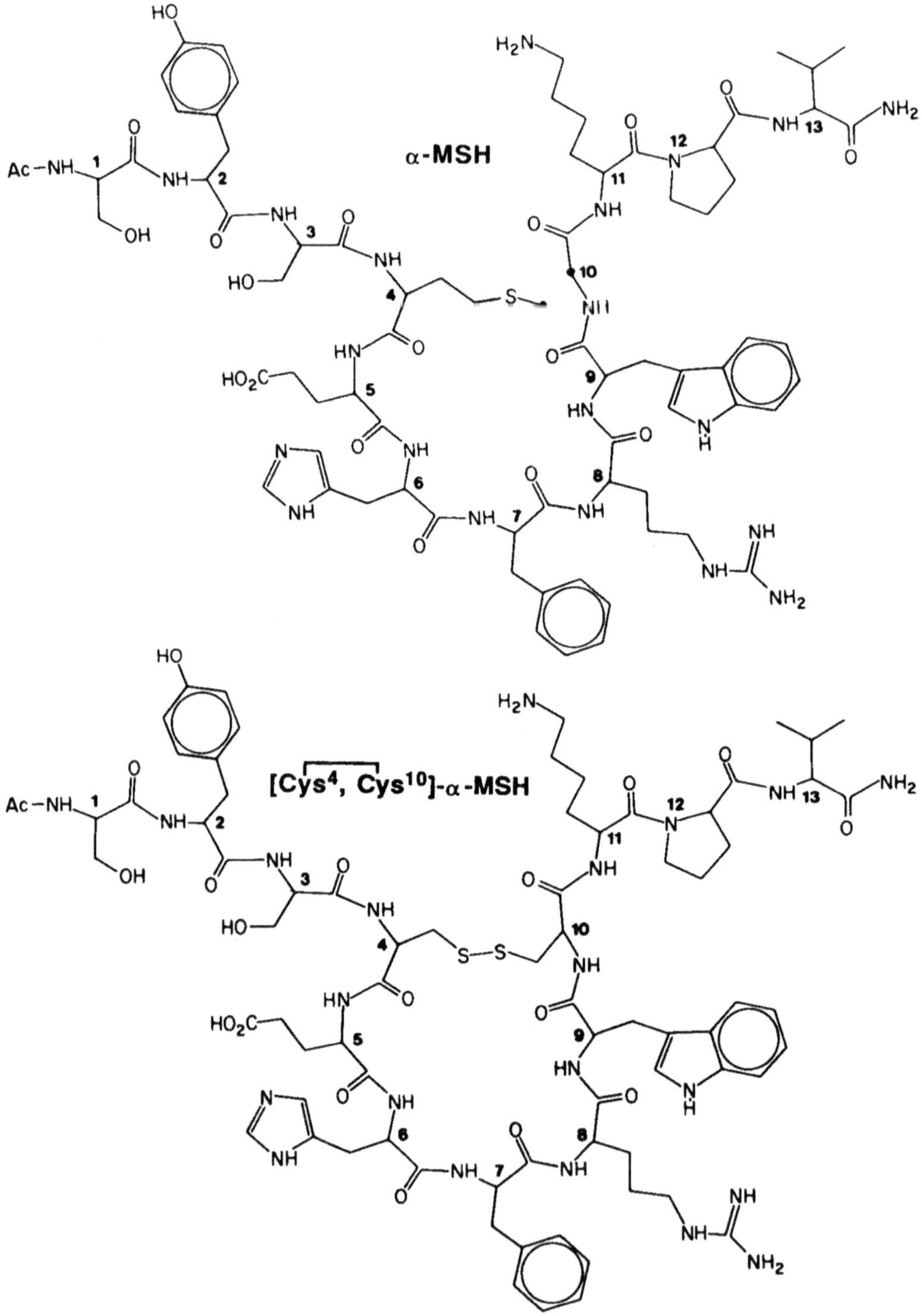

FIGURE 6. Comparative chemical structures of α-MSH and [Cys[4],Cys[10]]α-MSH.

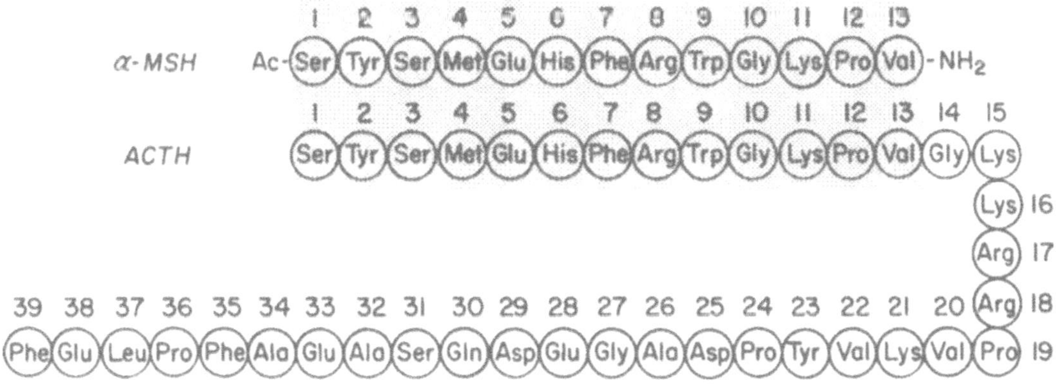

FIGURE 7. Comparative primary structure of α-MSH and ACTH.

Table 6
REPORTED HETEROSPECIFIC MODULATORS OF MSH BIOACTIVITY

	Tissue system	Ligand	Bioactivity
I.	Lizard (*A. carolinensis*) skin melanophores	Norepinephrine or epinephrine	Inhibition of MSH-stimulated melanosome dispersion[60]
		Norepinephrine or dopamine	Inhibition of [*p*-azido-Phe[13]]α-MSH-stimulated melanosome dispersion but not photoaffinity labeling of MSH receptors[61]
II.	Frog (*R. pipiens*) skin melanophores	Norepinephrine or melatonin	Inhibition of MSH-stimulated melanosome dispersion and cAMP production[62-64]

REFERENCES

1. **Ariens, E. J., Beld, A. J., Rodrigies de Miranda, J. F., and Simonis, A. M.,** The pharmacon-receptor-effector concept: a basis for understanding the transmission of information in biological system, in *The Receptors: General Principles and Procedures*, Vol. 1, O'Brien, R. D. Ed., Plenum Press, New York, 1979, 33.
2. **Catt, K. J., Harwood, J. P., Aguilera, G., and Dufau, M. L.,** Hormonal regulation of peptide receptors and target cell responses, *Nature (London)*, 280, 109, 1979.
3. **Rodbell, M.,** The role of hormone receptors and GTP-regulatory proteins in membrane transduction, *Nature (London)*, 284, 1722, 1980.
4. **Hirata, F. and Axelrod, J.,** Phospholipid methylation and biological signal transmission, *Science*, 209, 1082, 1980.
5. **Means, A. R. and Dedman, J. R.,** Calmodulin in endocrine cells and its multiple roles in hormone action, *Mol. Cell. Endocrinol.*, 19, 215, 1980.
6. **Ross, E. M. and Gilman, A. G.,** Biochemical properties of hormone-sensitive adenylate cyclase, *Ann. Rev. Biochem.*, 49, 533, 1980.
7. **Nishizuka, Y.,** Turnover of inositol phospholipids and signal transduction, *Science*, 225, 1365, 1984.
8. **Berridge, M. J. and Irvine, R. F.,** Inositol trisphosphate, a novel second messenger in cellular signal transduction, *Nature (London)*, 312, 315, 1984.
9. **Schramm, M. and Selinger, S.,** Message transmission: receptor controlled adenylate cyclase system, *Science*, 225, 1350, 1984.

10. **Hollenberg, M. D.,** Examples of homospecific and heterospecific receptor regulation, *Trends Pharmacol., Sci.*, 242, 1985.
11. **Hollenberg, M. D.,** Biochemical mechanisms of receptor regulation, *Trends Pharmacol., Sci.*, 299, 1985.
12. **Hollenberg, M. D.,** Pathophysiological and therapeutic implications of receptor regulation, *Trends Pharmacol., Sci.*, 3334, 1985.
13. **Langley, J. N.,** On nerve endings and on special excitable substances, *Proc. R. Soc. London Ser. B.*, 78, 170, 1906.
14. **Ehrlich, P.,** On partial functions of the cell (1908 Nobel Lecture), *in Collected Papers of P. Ehrlich*, Vol. 3, Himmelweit, F., Marquardt, M., and Dale, H., Eds., Pergamon Press, Oxford, 1956, 183.
15. **Clark, A. J.,** The reaction between acetylcholine and muscle cells, *J. Physiol. (London)*, 61, 530, 1926.
16. **Hruby, V. J.,** Structure and conformation related to activity of peptide hormones, in *Perspectives in Peptide Chemistry*, Eberle, A., Geiger, R., and Weiland, T., Eds., S. Karger, Basel, 1981, 209.
17. **Varga, J. M., Dipasquale, A., Pawelek, J., McGuire, J. S., and Lerner, A. B.,** Regulation of melanocyte stimulating level: discontinuous binding of hormone to synchronized mouse melanoma cells during the cell cycle, *Proc. Natl. Acad. Sci. U.S.A.*, 71, 1590, 1974.
18. **Eberle, A., Kriwaczek, V. M., and Schwyzer, R.,** Hormone-receptor interactions: melanotropic activities of covalent serum albumin complexes with α-melanotropin, α-melanotropin fragments and enkephalin, *Fed. Eur. Biochem. Soc. Lett.*, 80, 246, 1977.
19. **DiPasquale, A., Varga, J. M., Moellman, G., and McGuire, J.,** Synthesis of a hormonally active conjugate of α-MSH, ferretin, and fluorescein, *Anal. Biochem.*, 84, 37, 1978.
20. **Eberle, A. N., Kriwaczek, V. M., and Schulster, D.,** Corticotropin and melanotropin: organization and transmission of hormonal information, in *Perspectives in Peptide Chemistry*, Eberle, A., Geiger, R., and Weiland, T., Eds., S., Karger, Basel, 1981, 407.
21. **Kriwaczek, V. M., Eberle, A., Muller, M., and Schwyzer, R.,** Tobacco mosaic virus as a carrier for small molecules. I. The preparation and characterization of TMV/α-melanotropin conjugate, *Helv. Chim. Acta*, 61, 1232, 1978.
22. **Schwyzer, R.,** Studies on peptide hormone/receptor interactions, in *Peptides, Structure and Biological Function*, Gross, E. and Meienhofer, J., Eds., Pierce Chemical Company, Rockford, Ill., 1979, 993.
23. **Sawyer, T. K., Sanfilippo, P. J., Hruby, V. J., Engel, M. H., Heward, C. B., Burnett, J. B., and Hadley, M. E.,** 4-Norleucine,7-*D*-phenylalanine-α-melanocyte-stimulating hormone: a highly potent α-melanotropin with ultralong biological activity, *Proc. Natl. Acad. Sci. U.S.A.*, 77, 5754, 1980.
24. **Eberle, A. N.,** Photoaffinity labelling of MSH receptors of *Anolis* melanophores: irradiation technique and MSH photolabels for irreversible stimulation, *J. Receptor Res.*, 4, 315, 1984.
25. **Horowitz, J. M., Mikuckis, G. M., and Longshore, M. A.,** The response of single melanophores to extracellular and intracellular iontophoretic injection of melanocyte-stimulating hormone, *Endocrinology*, 106, 770, 1980.
26. **Bost, K. L. and Blalock, J. E.,** Molecular characterization of a corticotropin (ACTH) receptor, *Mol. Cell. Endocrinol.*, 44, 1, 1986.
27. **Medzihradszky, K.,** Synthesis and biological activity of adrenocorticotropic and melanotropic hormones, in *Recent Developments in the Chemistry of Natural Carbon Compounds*, Bogner, K., Bruckner, R., and Szantay, C., Eds., Hungarian Academy of Sciences, Budapest, 1976, 207.
28. **Eberle, A. N., Kriwaczek, V. M., and Schwyzer, R.,** Studies on tyrosinase stimulation, binding and degradation of α-MSH interacting with non-synchronized mouse melanoma cells in culture, in *Peptides, Structure and Biological Function*, Gross, E. and Meienhofer, J., Eds., Pierce Chemical Company, Rockford, Ill., 1979, 1033.
29. **Wilkes, B. C., Hruby, V. J., Yamamura, H. I., Akiyama, K., de L. Castrucci, A. M., Hadley, M. E., Andrews, J. R., and Wan, Y.-P.,** Synthesis of tritium labeled Ac-[Nle^4,D-Phe^7]-α-MSH_{4-11}-NH_2: a superpotent melanotropin with prolonged biological activity, *Life Sci.*, 34, 977, 1984.
30. **Lambert, D. T., Stachelek, C., Varga, J. M., and Lerner, A. B.,** Iodination of β-MSH, *J. Biol. Chem.*, 257, 8211, 1982.
31. **Eberle, A. and Hubscher, W.,** α-Melanotropin labelled at its $tyrosine^2$ residue: synthesis and biological activities of 3′-$iodotyrosine^2$-, 3′-^{125}$iodotyrosine^2$-, 3′,5′-$diiodotyrosine^4$-, and (3′,5′-$^3H_2$)$tyrosine^2$-α-melnotropin, and of related peptides, *Helv. Chim. Acta*, 62, 2460, 1979.
32. **Heward, C. B., Kreutzfeld, K. L., Hadley, M. E., Larson, B., Sawyer, T. K., and Hruby, V. J.,** Preparation of a radiolabeled melanotropin suitable for use as a tracer in a radioreceptor assay: [^{125}I-Tyr^2,Nle^4]-α-melanotropin, *Pigment Cell*, 6, 339, 1981.
33. **Ghanem, G. and Libert, A.,** Alpha-melanocyte stimulating hormone receptors on human malignant melanoma cells in culture, (Abstr., *13th Intl. Pigment Cell Conf.,*) *J. Invest. Dermatol.*, 87, 414, 1986.
34. **Abe, K., Butcher, R. W., Nicholson, W. E., Bairde, C. E., Liddle, R. A., and Liddle, G. W.,** Adenosine 3′,5′-monophosphate (cyclic AMP) as the mediator of melanocyte stimulating hormone (MSH) and norepinephrine on the frog skin, *Endocrinology*, 84, 362, 1969.

35. **Wong, G., Pawelek, J., Sansone, M., and Morowitz, J.,** Response of mouse melanoma cells to melanocyte stimulating hormone, *Nature (London)*, 248, 351, 1974.
36. **O'Keefe, E. and Cuatrecasas, P.,** Cholera toxin mimics melanocyte stimulating hormone in inducing differentiation in melanoma cells, *Proc. Natl. Acad. Sci. U.S.A.*, 71, 2500, 1974.
37. **Legros, F., Coel, J., Doyen, A., Hanson, P., van Tiegham, N., Vercammem-Grandjean, A., Fruhling, J., and Lejeune, F. J.,** α-Melanocyte-stimulating hormone binding and biological activity in a human melanoma cell line, *Cancer Res.*, 41, 1539, 1981.
38. **Kreiner, P. W., Gold, C. J., Keirns, J. J., Brock, W. H., and Bitensky, M. W.,** MSH-sensitive adenylyl cyclase in the Cloudman melanoma, *Yale J. Biol. Med.*, 46, 583, 1973.
39. **Bregman, M. D., Sawyer, T. K., Hadley, M. E., and Hruby, V. J.,** Adenosine and divalent cation effects on S-91 melanoma adenylate cyclase, *Arch. Biochem. Biophys.*, 201, 1, 1980.
40. **Bitensky, M. W. and Burnstein, S. B.,** Effects of cyclic adenosine monophosphate and melanocyte-stimulating hormone on frog skin in vitro, *Nature (London)*, 208, 1282, 1965.
41. **Novales, R. R. and David, W. J.,** Melanin dispersing effect of adenosine 3′,5′-monophosphate on amphibian melanocytes, *Endocrinology*, 81, 283, 1967.
42. **Hadley, M. E. and Goldman, J. M.,** Effects of cyclic 3′,5′-AMP and other adenine nucleotides on the melanophores of the lizard (Anolis carolinensis), *Br. J. Pharmacol.*, 37, 650, 1969.
43. **Goldman, J. M. and Hadley, M. E.,** Evidence for separate receptors for melanophore stimulating hormone and catecholamine regulation of cyclic AMP in the control of melanophore responses, *Br. J. Pharmacol.*, 39, 160, 1970.
44. **Fuller, B. B. and Hadley, M. E.,** Transcriptional and translational requirements for MSH stimulation of tyrosinase in melanoma cells, *Pigment Cell*, 4, 97, 1979.
45. **Hruby, V. J., Wilkes, B. C., Cody, W. L., Sawyer, T. K., and Hadley, M. E.,** Melanotropins: structural, conformational and biological considerations in the development of superpotent and superprolonged analogs, *Peptide Protein Rev.*, 3, 1, 1984.
46. **Hadley, M. E., Anderson, B., Heward, C. B., Sawyer, T. K., and Hruby, V. J.,** Calcium-dependent prolonged effects on melanophores of [4-norleucine,7-D-phenylalanine]-α-melanotropin, *Science*, 213, 1025, 1981.
47. **Veseley, D. L. and Hadley, M. E.,** Ionic requirements for melanophore-stimulating hormone (MSH) action of melanophres, *Comp. Biochem. Physiol.*, 62A, 501, 1979.
48. **Hadley, M. E., Heward, C. B., Hruby, V. J., Sawyer, T. K., and Yang, Y. C. S.,** Biological actions of melanocyte-stimulating hormone, in *Peptides of the Pars Intermedia (Ciba Foundation Symp. 81)*, Pitman Medical, London, 1981, 244.
49. **Sawyer, T. K.,** Design and Synthesis of Structural, Steroisomeric and Conformationally-Restricted Analogues of α-Melanotropin: Comparative Biological Properties of Melanophores and Melanoma Cells, Ph.D. dissertation, University of Arizona, Tucson, 1981.
50. **de Graan, P. N. E., van de Kamp, A. J., Hup, D. R. W., Gispen, W. H., and van de Veerdonk, F. C. G.,** Calcium requirement for α-MSH action on melanophores: studies with forskolin, *J. Receptor Res.*, 4, 521, 1984.
51. **Eberle, A. N.,** MSH receptors, in *Cellular Receptors for Hormones and Neurotransmitters*, Schulster, D. and Levitzki, A., Eds., John Wiley & Sons, New York, 1980, 219.
52. **Schwyzer, R. and Eberle, A.,** On the molecular mechanisms of α-MSH receptor interactions, *Front. Horm. Res.*, 4, 18, 1977.
53. **Sawyer, T. K., Hruby, V. J., Hadley, M. E., and Engel, M. H.,** α-Melanocyte stimulating hormone: chemical nature and mechanism of action, *Am. Zool.*, 23, 529, 1983.
54. **Wunderlin, R., Sharma, S. D., Minakakis, P., and Schwyzer, R.,** Melanotropin receptors. II. Synthesis and biological activity of α-melanotropin/tobacco mosiac virus disulfide conjugate, *Helv. Chim. Acta*, 68, 12, 1985.
55. **Lucas, A. M., Shuster, S. and Thody, A. J.,** Modulation of melansome dispersion by protein kinase C, *J. Invest. Dermatol.*, (Abstr. *Proc. 13th Intl. Pigment Cell Conf.)*, 87, 416, 1986.
56. **Hadley, M. E.,** Calcium-dependent irreversible effect of ionophore A23187 on melanophores, *Pigment Cell Res.*, 1986, 41, 57, 1987.
57. **Sargent, D. F. and Schwyzer, R.,** Membrane lipid phase as catalyst for peptide-receptor interactions, *Proc. Natl. Acad. Sci. U.S.A.*, 83, 5774, 1986.
58. **Sawyer, T. K., Hruby, V. J., Darman, P. S., and Hadley, M. E.,** [Half-Cys^4,half-Cys^{10}]-α-melanocyte-stimulating hormone: a cyclic α-melanotropin exhibiting superagonist biological activity, *Proc. Natl. Acad. Sci. U.S.A.*, 79, 1751, 1982.
59. **Vinson, G. P., Whitehouse, B. J., Bateman, A., Hruby, V. J., Sawyer, T. K., and Darman, P. S.,** α-MSH analogues and adrenal glomerulosa function, *Life Sci.*, 35, 603, 1984.
60. **Sawyer, T. K., Staples, D. J., Affholter, J., Al-Obeidi, F. A., Hruby, V. J., and Hadley, M. E.,** [D-phenylalanine-7]-substituted MSH active site tri- and tetrapeptides exhibiting high melanotropic potency and prolonged biological activity, 1987, manuscript in preparation.

61. **Cheitlin, R., Buckley, D. I., and Ramachandran, J.,** The role of extracellular calcium in corticotropin-stimulated steriodogenesis, *J. Biol. Chem.*, 260, 5323, 1985.
62. **Kojima, I., Kojima, K., and Rasmussen, H.,** Role of calcium and cAMP in the action of adrenocorticotropin on aldosterone secretion, *J. Biol. Chem.*, 260, 4248, 1985.
63. **Goldman, J. M. and Hadley, M. E.,** *In vitro* demonstration of adrenergic receptors controlling melanophore responses in the lizard, *Anolis carolinensis, J. Pharmacol. Exp. Ther.*, 166, 1, 1969.
64. **Eberle, A. N. and Girard, J.,** Photoaffinity labelling of MSH receptors on *Anolis* melanophores: effects of catecholamines, calcium and forskolin, *J. Receptor Res.*, 5, 59, 1985.
65. **Abe, K., Robison, G. A., Liddle, G. W., Butcher, R. W., Nicholson, W. E., and Baird, C. E.,** Role of cyclic AMP in mediating the effects of MSH, norepinephrine, and melatonin on frog skin color, *Endocrinology*, 85, 674, 1969.
66. **Heward, C. B. and Hadley, M. E.,** Structure-activity relationships of melatonin and related indoleamines, *Life Sci.*, 17, 1167, 1975.
67. **Wilkes, B. C., Hruby, V. J., de L. Castrucci, A. M., Sherbrooke, W. C., and Hadley, M. E.,** Synthesis of a cyclic melanotropic peptide exhibiting both melanin-concentrating and -dispersing activities, *Science*, 224, 1111, 1984.
68. **Hadley, M. E., Zechel, C., Wilkes, B. C., de L. Castrucci, A. M., Visconti, M. A., Pozo-Alonzo, M., and Hruby, V. J.,** Differential structural requirements for MSH and MCH activities of melanin concentrating hormone, *Life Sci.*, 40, 1845, 1987.
69. **de L. Castrucci, A. M., Hadley, M. E., and Hruby, V. J.,** Melanocyte stimulating hormone (MSH) and melanin concentrating hormone (MCH) exhibit both MSH-like and MCH-like activity depending on the presence or absence of extracellular calcium, *Proc. Natl. Acad. Sci. U.S.A.*, 1987, submitted.

Chapter 6

CYCLIC, CONFORMATIONALLY CONSTRAINED MELANOTROPIN ANALOGS: STRUCTURE-FUNCTION AND CONFORMATIONAL RELATIONSHIPS

Wayne L. Cody, Mac E. Hadley, and Victor J. Hruby

TABLE OF CONTENTS

I. Introduction 76

II. Early Biological Activities and Structural Analyses of Linear and Cyclic Melanotropins 76
- A. Biological Properties of Melanotropins 76
- B. Structure-Function Studies 77
  - 1. Importance of the Seven Position to Biological Potency 77
  - 2. Design and Preparation of $[\overline{Cys^4,Cys^{10}}]\alpha$-MSH 78

III. Analogs of α-MSH Cyclized Via a Cysteine Disulfide Bridge in the Four and Ten Positions 80
- A. Importance of N-Terminal Amino Acids 80
  - 1. Linear Analogs 80
  - 2. Cyclic Analogs 80
- B. Importance of the C-Terminal Amino Acids 81
  - 1. Linear Analogs 81
  - 2. Cyclic Analogs 81
- C. Importance of the D-Phenylalanine Substitution in Cyclic Analogs of α-MSH 83
  - 1. Linear Analogs 83
  - 2. Cyclic Analogs 83

IV. Analogs of α-MSH Cyclized Via a Disulfide Bridge in the Four and Ten Positions with One or Both of the Cysteines Replaced by Other Sulfur-Containing Amino Acids or Derivatives 83
- A. Substitutions for Cysteine in the Four Position 84
  - 1. Ac-$[\overline{\text{D-}Cys^4,Cys^{10}}]\alpha$-$MSH_{4\text{-}13}$-$NH_2$ 85
  - 2. $[\overline{Mpa^4,Cys^{10}}]\alpha$-$MSH_{4\text{-}13}$-$NH_2$ 86
  - 3. $[\overline{Mpa^4,Cys^{10}}]\alpha$-$MSH_{4\text{-}13}$-$NH_2$ 86
  - 4. Ac-$[\overline{Hcy^4,Cys^{10}}]\alpha$-$MSH_{4\text{-}13}$-$NH_2$ 86
- B. Substitution of Penicillamine for Cysteine in the Four and/or Ten Positions 87

V. Miscellaneous Cyclic Analogs 87
- A. Carba Analogs of Ac-$[\overline{Cys^4,Cys^{10}}]\alpha$-$MSH_{4\text{-}13}$-$NH_2$ 88
  - 1. Oxidized Carba Analogs 88
- B. Cyclic Analogs Chemically Reduced to the Linear Dithiol Analogs 89

VI. Other Biological Activities of Cyclic Melanotropins 89

VII. Summary and Future Perspectives 89

Acknowledgments 90

References 90

I. INTRODUCTION

During the past quarter century, significant effort has been directed at determining the key residues and the three-dimensional structure which are required for the biological activity of linear melanotropins. The designation melanotropin has been applied to a large number of structurally similar peptides (e.g., α-melanocyte-stimulating hormone, [α-MSH], β-MSH, and γ-MSH). It has been suggested that all the melanotropins may be evolutionarily related, but α-MSH, [(Ac-Ser^{1}-Tyr^{2}-Ser^{3}-Met^{4}-Glu^{5}-His^{6}-Phe^{7}-Arg^{8}-Trp^{9}-Gly^{10}-Lys^{11}-Pro^{12}-Val^{13}-NH_2), a linear tridecapeptide containing the first 13 amino acids of adrenocorticotropin (ACTH)$^{1\text{-}39}$, is generally considered to be the physiologically relevant melanotropin[1,2] and will be the primary focus of this chapter. The results of many of these structure-function studies on α-MSH have been critically reviewed in a previous chapter in this series (Chapter 6, this Volume).

Until recently, most structure-function studies were based upon individual amino acid substitutions or deletions in α-MSH and/or shorter linear fragments. A careful examination of the structure-activity relationships in these linear analogs and application of rational design techniques led to the preparation of a series of cyclic analogs of α-MSH, many of which possess greater biological potency than their linear counterparts in the classical in vitro frog skin bioassay. Since cyclization of a peptide generally leads to considerable conformational restriction of the peptide backbone and, to a lesser degree, the side chains, a critical examination of these cyclic analogs can provide important insight into the possible three-dimensional, biologically active topology of α-MSH. These topologies are currently under investigation by examination of the structure-biological activity relationships in conjunction with modern biophysical methods (e.g., circular dichroism [CD], nuclear magnetic resonance [NMR], computational methods, etc.) Prior to a discussion of the structure-activity relationships in cyclic melanotropins and their inferences for further rational design of melanotropins, the structure-activity relationships of relevant selected linear melanotropins will be briefly reviewed.

II. EARLY BIOLOGICAL ACTIVITIES AND STRUCTURAL ANALYSES OF LINEAR AND CYCLIC MELANOTROPINS

A. Biological Properties of Melanotropins

α-MSH is primarily known for its ability to stimulate integumental melanocytes,[1,2] but has also been implicated in a variety of physiological and neurological processes.[3-6] The pigmentary activity of α-MSH has been examined primarily in the in vitro frog (*Rana pipiens*) and lizard (*Anolis carolinensis*) skin bioassays and, to a lesser degree, in the corresponding in vivo bioassays. The in vitro biological assays allow quantitative measurements of the potency of melanotropins over a large dosage range (10^{-4} to 10^{-13} *M*).[7] Two further bioassays commonly utilized for the measurement of the potency of melanotropins are the adenylate cyclase[8] and the tyrosinase[9,10] bioassays.

Table 1
RELATIVE IN VITRO POTENCIES FOR LINEAR C-TERMINAL FRAGMENTS OF α- MSH

Peptide	Biological activities[a] Frog	Lizard	Ref.
α-MSH	1.0	1.0	
[des-Ac]α-$MSH_{1\text{-}13}$-NH_2	0.10	N.D.	12
Ac-α-$MSH_{2\text{-}13}$-NH_2	0.50	N.D.	12, 23
Ac-α-$MSH_{3\text{-}13}$-NH_2	0.20	N.D.	12, 25
Ac-α-$MSH_{4\text{-}13}$-NH_2	0.10	N.D.	12, 23
Ac-α-$MSH_{5\text{-}13}$-NH_2	0.012	N.D.	12, 23
[Nle^4]α-MSH	2.2	1.5	12, 17
[Nle^4,D-Phe^7]α-MSH	60.0	5.0	11, 12

[a] Biological potencies are measured relative to α-MSH (1.0) over the linear portion of the dose-response curve.

Table 2
D-L RATIOS OF AMINO ACID RESIDUES OF α-MSH AFTER HEAT-ALKALI TREATMENT[a]

Amino acid residue	Percent D-amino acid residue
Ser-1,3	52.0[b]
Tyr-2	16.0
Met-4	24.0
Glu-5	14.5
His-6	32.5
Phe 7	30.9
Arg-8	28.0
Trp-9	9.0
Lys-11	1.4
Pro-12	0.4
Val-13	3.9

[a] 100°C, 10 min in 0.1 *N* aqueous NaOH.
[b] Values represent the average of the sum for the serine residues in positions 1 and 3.

B. Structure-Function Studies

1. Importance of the Seven Position to Biological Potency

Earlier structure-function studies of α-MSH have shown that the stereochemistry of the amino acid in the seven position of α-MSH (phenylalanine) has dramatic effects on the potency of the peptide (Table 1).[11-14] It was shown that heat-alkali treatment of α-MSH produced a complex mixture of peptides which showed retardation of the biological response, but with potentiation and prolongation of the response of the in vitro frog skin bioassay.[12,15] Prolonged activity was defined as an increase of the duration of the biological response (skin darkening) *after* removal of the peptide from the assay medium (Ringer solution) and bathing the frog (or lizard) skin. The amount of racemization of individual amino acids in α-MSH was quantitated by gas chromatography (Table 2).[16] This study revealed that the amino acids in the 6-8 sequence (-His-Phe-Arg) were racemized to the greatest extent. On the basis of these data, model building, Chou-Fasman calculations, and conformational considerations related to other structure-function studies of α-MSH, the phenylalanine in the seven position was replaced by its optical enantiomer, D-phenylalanine. The resulting analog, [Nle^4,D-Phe^7]α-MSH,[11,12] was more potent than the native hormone (60- and 5-fold in the in vitro frog and lizard skin bioassay, respectively, Table 1) and also possessed prolongation of the biological response. Norleucine (Nle) replaced the methionine in position four of α-MSH because it is pseudoisosteric to methionine, but does not contain sulfur which is known to oxidize easily with loss of potency in α-MSH.[17]

2. Design and Preparation of [Cys^4,Cys^{10}]α-MSH

Based upon the enhanced potency of [Nle^4,D-Phe^7]α-MSH, the ability of D-amino acids to stabilize reverse turns, and application of rational design techniques, a cyclic analog of α-MSH was prepared ([Cys^4,Cys^{10}]α-MSH).[18,19] Careful examination of α-MSH revealed that cyclization between the four and ten positions would maintain the residues believed critical to melanotropic activity and would keep Phe-7 at the center of a reverse turn. Cyclization was performed by the pseudoisosteric substitution of methionine in the four position and glycine in the ten position with a cysteine bridge (Figure 1). The resulting cyclic tridecapeptide, [Cys^4,Cys^{10}]α-MSH, was more than an order of magnitude (≥ tenfold) more potent than α-MSH in the in vitro frog skin bioassay over the linear portion of the dose-response curve and at least two times as potent as the native hormone in the in vitro lizard skin bioassay.[18,19] Prolonged exposure of the frog skins to the cyclic analog, however, showed that it had a minimum effective dose potency of 1000 to 10,000 times that of α-MSH in the former assay system. In neither of the bioassays did the cyclic molecule show any prolongation. This has been used as evidence to suggest that prolongation and potency are not correlated to the same biological event.[14,20,21]

The high potency of the cyclic tridecapeptide melanotropin, suggests several important ideas: (1) cyclization of α-MSH results in a conformation that can be readily recognized by the melanotropic receptor, (2) the preferred cyclic conformation may be very close to mimicking the ideal molecular substrate for the receptor, and (3) enhancement of potency and prolongation are not necessarily related to the same topological structural features. We will now begin a critical and systematic analysis of a series of cyclic analogs of α-MSH and their implications to the future design and preparation of highly potent melanotropin analogs.

Subsequent evaluation of structure-function relationships in the cyclic α-MSH series have demonstrated (Table 3) the following:

1. The relative importance of the C-terminal (-Lys-Pro-Val-) but not the N-terminal (-Ser-Tyr-Ser-) tripeptide sequences to the potency of cyclic Cys^4,Cys^{10} disulfide-containing analogs with no modifications of the amino acids in the five through nine sequence.
2. The importance of the C-terminal tripeptide sequences to the potency of cyclic Cys^4,Cys^{10} disulfide-linked analogs with the D-phenylalanine substitution for the L-phenylalanine in the seven position.
3. Potent analogs cyclized between the four and ten positions via a disulfide bridge can be prepared, where one or both of the cysteines are replaced by other sulfur-containing amino acids or related compounds.
4. Other cyclic analogs, including "carba" analogs and oxidized "carba" analogs, generally have reduced potency.

In general, the relative potencies of individual cyclic α-MSH analogs have been assessed and compared in both the in vitro frog and lizard skin bioassay.[12,22] (A minimum of two

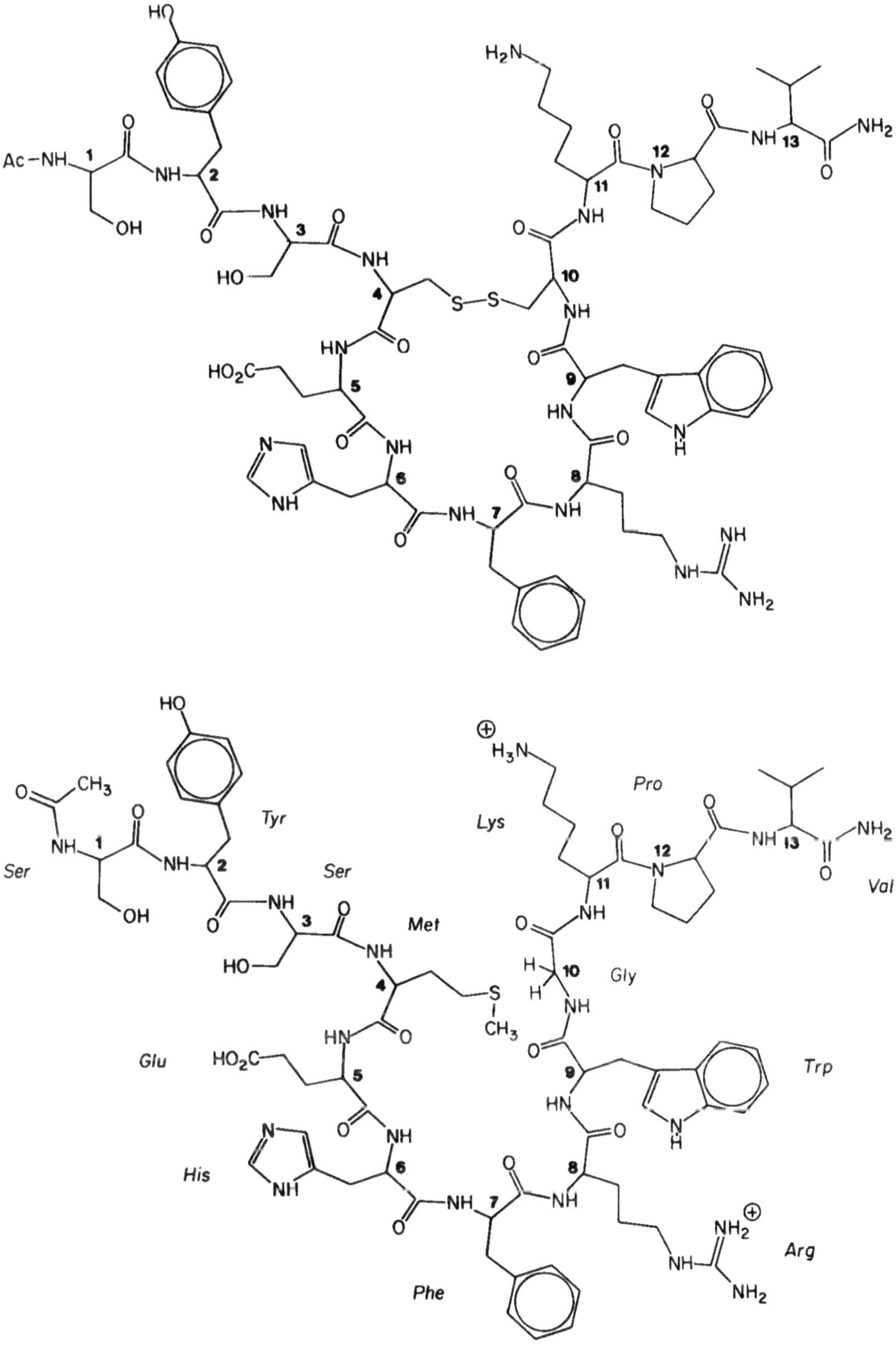

FIGURE 1. An illustration of the pseudoisosteric substitution of cysteine in the four and ten positions: α-MSH (top) and the pseudoisosteric $[\overline{\text{Cys}^4,\text{Cys}^{10}}]$α-MSH (bottom).

Table 3
RELATIVE IN VITRO POTENCIES FOR CYCLIC DISULFIDE-LINKED MELANOTROPINS ([CYS⁴, CYS¹⁰] ANALOGS)

Peptide	Biological activities[a] Frog	Lizard	Ref.
α-MSH	1.00	1.00	
[Cys^4,Cys^{10}]α-MSH[b]	20.0	2.0	12, 18, 19
Ac-[Cys^4,Cys^{10}]α-MSH_{4-13}-NH_2	30.0	0.60	12, 27, 30
Ac-[Cys^4,Cys^{10}]α-MSH_{4-12}-NH_2	10.0	1.5	14, 30
Ac-[Cys^4,Cys^{10}]α-MSH_{4-11}-NH_2	0.16	0.07	12, 27, 30
Ac-[Cys^4,Cys^{10}]α-MSH_{4-10}-NH_2	0.06	0.003	12, 27, 30
Ac-[Cys^4,D-Phe^7,Cys^{10}]α-MSH_{4-13}-NH_2	6.0	6.0	14
Ac-[Cys^4,D-Phe^7,Cys^{10}]α-MSH_{4-12}-NH_2	20.0	6.0	27, 30
Ac-[Cys^4,D-Phe^7,Cys^{10}]α-MSH_{4-11}-NH_2	2.5	3.0	14
Ac-[Cys^4,D-Phe^7,Cys^{10}]α-MSH_{4-10}-NH_2	0.75	0.5	14

[a] Biological potencies are measured relative to α-MSH (1.0) over the linear potion of the dose-response curve.

[b] The potency of the minimum effective dose is 1000 to 10,000 times greater than α-MSH.

different species bioassays is used because different melanotropin receptors have different structural requirements).[12,22]

III. ANALOGS OF α-MSH CYCLIZED VIA A CYSTEINE DISULFIDE BRIDGE IN THE FOUR AND TEN POSITIONS

A. Importance of the N-Terminal Amino Acids

1. Linear Analogs

The contribution of the N-terminal tripeptide (-Ser-Tyr-Ser-) in α-MSH to the overall potency of the molecule has been examined extensively in linear analogs (Table 1).[12,23,24] Removal of individual amino acids from the N-terminus of α-MSH resulted in a slight loss of biological potency in the in vitro frog skin (*R. pipiens*) bioassay. The 2-13[12,13] and 3-13[12,25] sequences were two- and fivefold less potent, respectively, than the parent molecule. Loss of either the N-terminal acetyl group[12,26] or the tripeptide (-Ser-Tyr-Ser-)[12] led to a corresponding tenfold decrease in biological potency. It was concluded that the N-terminal tripeptide (-Ser-Tyr-Ser-) was not a highly important sequence to the overall potency of the molecule. Similar results have been obtained in both the in vitro frog and lizard skin bioassays for the Nle^4-substituted series (Tables 1 and 4).[24]

2. Cyclic Analogs

Examination of analogs and fragments of α-MSH cyclized via a cysteine bridge in the four and ten positions led to similar results. In the tridecapeptide, cystine cyclization led to an analog [Cys^4,Cys^{10}]α-MSH) that was significantly more potent than the corresponding linear sequence (10- to 20-fold in the linear portion of the dose-response curve) in the in vitro frog skin bioassay, but was only slightly more potent than α-MSH (twofold) in the in vitro lizard skin bioassay. Interestingly, removal of the N-terminal tripeptide sequence, (-Ser-Tyr-Ser-) from [Cys^4,Cys^{10}]α-MSH, resulted in an analog, Ac-[Cys^4,Cys^{10}]α-MSH_{4-13}-NH_2[12,27] that was slightly more potent than the cyclic tridecapeptide in the frog skin bioassay, but was slightly less potent in the lizard skin bioassay. The data from these two

Table 4
RELATIVE IN VITRO POTENCIES FOR LINEAR N-TERMINAL FRAGMENTS OF α-MSH

Peptide	Biological activities[a] Frog	Lizard	Ref.
α-MSH	1.0	1.0	—
[Nle[4]]α-MSH	2.2	1.5	12, 17
Ac-[Nle[4]]α-MSH_{4-13}-NH_2	1.0	2.0	12, 17, 24
Ac-[Nle[4]]α-MSH_{4-12}-NH_2	0.10	5.0	24
Ac-[Nle[4]]α-MSH_{4-11}-NH_2	0.002	1.0	12, 29
Ac-[Nle[4]]α-MSH_{4-10}-NH_2	0.0002	0.06	12, 28
Ac-[Nle[4],D-Phe[7]]α-MSH	60.0	5.0	11, 12
Ac-[Nle[4],D-Phe[7]]α-MSH_{4-11}-NH_2	0.16	8.0	12, 29
Ac-[Nle[4],D-Phe[7]]α-MSH_{4-10}-NH_2	0.02	10.0	12, 28

[a] Biological potencies are measured relative to α-MSH (1.0) over the linear portion of the dose-response curve.

cyclic analogs further illustrate the species differences of melanocyte receptors to the same melanotropic peptide. Similar to most cyclic analogs, Ac-[Cys[4],Cys[10]]α-$MSH_{4-13}NH_2$ was significantly more potent than the corresponding linear sequence (Figure 2, Tables 3 and 4).[27] The synthesis and corresponding biological properties of other N-terminal-shortened cyclic Cys[4],Cys[10] analogs have not been reported (e.g., the 2-13 or 3-13 sequences).

B. Importance of the C-Terminal Amino Acids

The contribution of the C-terminal tripeptide of α-MSH (-Lys-Pro-Val-) to the potency of α-MSH in linear peptide analogs also has been examined. In general, these studies were performed with norleucine substituted for methionine in the four position.[12,24,28,29] Removal of the C-terminal tripeptide (-Lys-Pro-Val-) in the Nle[4] series resulted in a 500-fold loss of potency on the classical in vitro frog skin bioassay (Table 4).[28] Restoration of lysine to the C-terminus of the 4-10 sequence did not enhance the potency;[29] however, incorporation of proline in the 12 position resulted in a 50-fold increase in potency with respect to Ac-[Nle[4]] α-MSH_{4-11}-NH_2.[24] The 4-12 nonapeptide analog possessed one tenth the potency of α-MSH or Ac-[Nle[4]]α-MSH_{4-13}-NH_2 on the frog skin bioassay (Table 4). Similar results were obtained in the in vitro lizard skin bioassay, although the potency differences were less dramatic.[12,17,24,28,29] These results suggest that Pro-12 is a critical residue for high melanotropic activity at the frog melanotropic receptor for linear melanotropins.

2. *Cyclic Analogs*

The biological properties of C-terminal shortened cyclic Cys[4],Cys[10] disulfide-linked analogs were similar to the linear analogs. For example, the 4-13 cyclic analog, Ac-[Cys[4],Cys[10]]α-MSH_{4-13}-NH_2, was more than 30-fold more potent than α-MSH in the frog skin and equipotent to α-MSH in the lizard skin bioassay (Table 3). Interestingly, removal of Val[13] resulted in a slight loss of potency in the frog skin, but a slight enhancement of potency in the lizard skin bioassay. The cyclic 4-11 analog was significantly less potent than the cyclic 4-12 analog in both bioassays.[12,30] These data illustrate the importance of Pro-12 in the cyclic, as well as the linear C-terminal shortened series to biological potency. The analog Ac-[Cys[4],Cys[10]]α-MSH_{4-10}-NH_2 was less potent than the cyclic 4-11 analog in both bioassays (~3- to 12-fold).[12,14,27,30] It has been suggested that this loss of potency may be due to the ability of lysine to provide a polar side chain which enhances aqueous solubility and thus affects the biological potency.

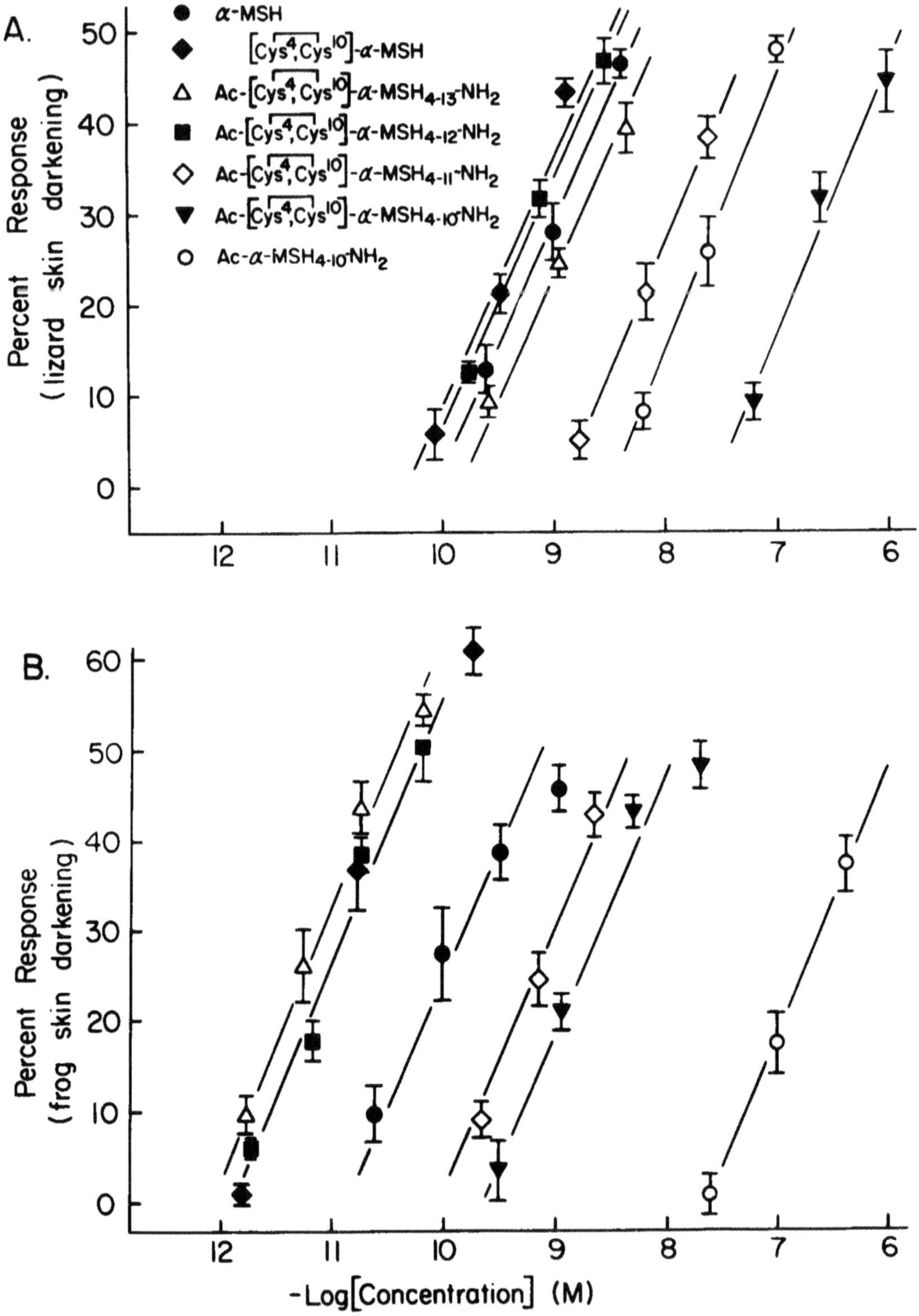

FIGURE 2. An illustration of the relative potencies of a series of cyclic cysteine-substituted C-terminal fragments of α-MSH in the (A) lizard and (B) frog skin bioassays.

A careful examination of the data reveals that the importance of the C-terminal tripeptide (-Lys-Pro-Val-) to the potency of cyclic melanotropins parallels the data for the linear analogs (Figure 2). In all cases, removal of amino acids from the C-terminus of α-MSH can have a large effect on bioactivity in either linear or cyclic melanotropins. Generally, cyclic ([Cys^4,Cys^{10}]) melanotropins are more potent than their linear counterparts (Met^4 or Nle^4). In the cyclic series, the 4-11 analog is more potent than the 4-10 analog, and the 4-11 cyclic analog is more potent than the Nle^4-4-11 linear analog; however, the 4-10 cyclic analog is less potent than the corresponding linear 4-10 analog (Ac-α-MSH$_{4-10}$-NH$_2$) in the lizard skin bioassay (Figure 2).[12,30] These results further illustrate the dramatic differences in biological response between frog and lizard skin melanocyte receptors to individual cyclic melanotropins and the structure-activity differences between linear and cyclic melanotropins.

In regards to prolongation of the melanotropic activity, the cyclic analogs were also similar to the linear analogs. None of the cyclic analogs or fragments of α-MSH, linear or cyclic, were prolonged with L-phenylalanine in the seven position.[12,19,27,30]

C. Importance of the D-Phenylalanine Substitution in Cyclic Analogs of α-MSH

1. Linear Analogs

As previously discussed, ([Nle^4,D-Phe^7]α-MSH was 5 to 60 times more potent than the corresponding L-Phe^7-containing analog in both in vitro bioassays. Similar results have been obtained for the D-Phe^7 substitution in shorter linear Nle^4-substituted fragments (i.e., 4-11 and 4-10).[12,28,29]

The D-Phe^7 substitution in the [Nle^4]α-MSH analog has also been shown to lead to a peptide with highly prolonged melanotropic activity.[11,12] Similar results were obtained for shorter fragments of the linear Nle^4-substituted series (i.e., 4-13, 4-12, and 4-11), except for the 4-10 analog which possessed no prolongation of the biological response in the frog skin bioassay.[12,28,29] This again shows that the biological properties of enhanced potency and prolongation are not necessarily related.

2. Cyclic Analogs

The D-Phe^7 substitution in the [Cys^4,Cys^{10}] series generally gave similar results.[14,30] Unfortunately, for comparison purposes, [Cys^4,D-Phe^7,Cys^{10}] α-MSH has never been prepared. The D-Phe^7-substitution in the 4-10 and 4-11 cysteine-bridged melanotropins led to analogs that were considerably more potent (Figure 3) than the corresponding L-Phe^7-containing analogs in both in vitro bioassays (10 to 200-fold) (Table 3).[14] However, the D-Phe^7 substitution did not lead to an enhancement of potency in either the 4-12 or 4-13 cyclic sequences (Table 3).[14,30] Both peptides were essentially equipotent to the corresponding cyclic L-Phe-containing cyclic fragment analogs.

The prolongation of melanotropic activity in the cyclic D-Phe^7 series differs significantly from that observed for the linear series. Both the cyclic [Cys^4,D-Phe^7, Cys^{10}] 4-12 and 4-13 analogs were prolonged, but the 4-10 and 4-11 analogs were not (Figure 4).[14,30] Interestingly, the linear 4-11 analog Ac-[Nle^4,D-Phe^7]α-MSH$_{4-11}$-NH$_2$ is prolonged (Figure 4). Since the phenomenon of prolongation is not completely understood, this observation may provide critical clues into unraveling the origins of prolonged activity.

IV. ANALOGS OF α-MSH CYCLIZED VIA A DISULFIDE BRIDGE IN THE FOUR AND TEN POSITIONS WITH ONE OR BOTH OF THE CYSTEINES REPLACED BY OTHER SULFUR-CONTAINING AMINO ACIDS OR DERIVATIVES

The high potency of the cyclic cysteine-bridged compounds has spurred considerable interest as to the conformational structure features of the molecule that lead to this enhanced potency. Since the molecule is conformationally restricted by virtue of the disulfide bridge,

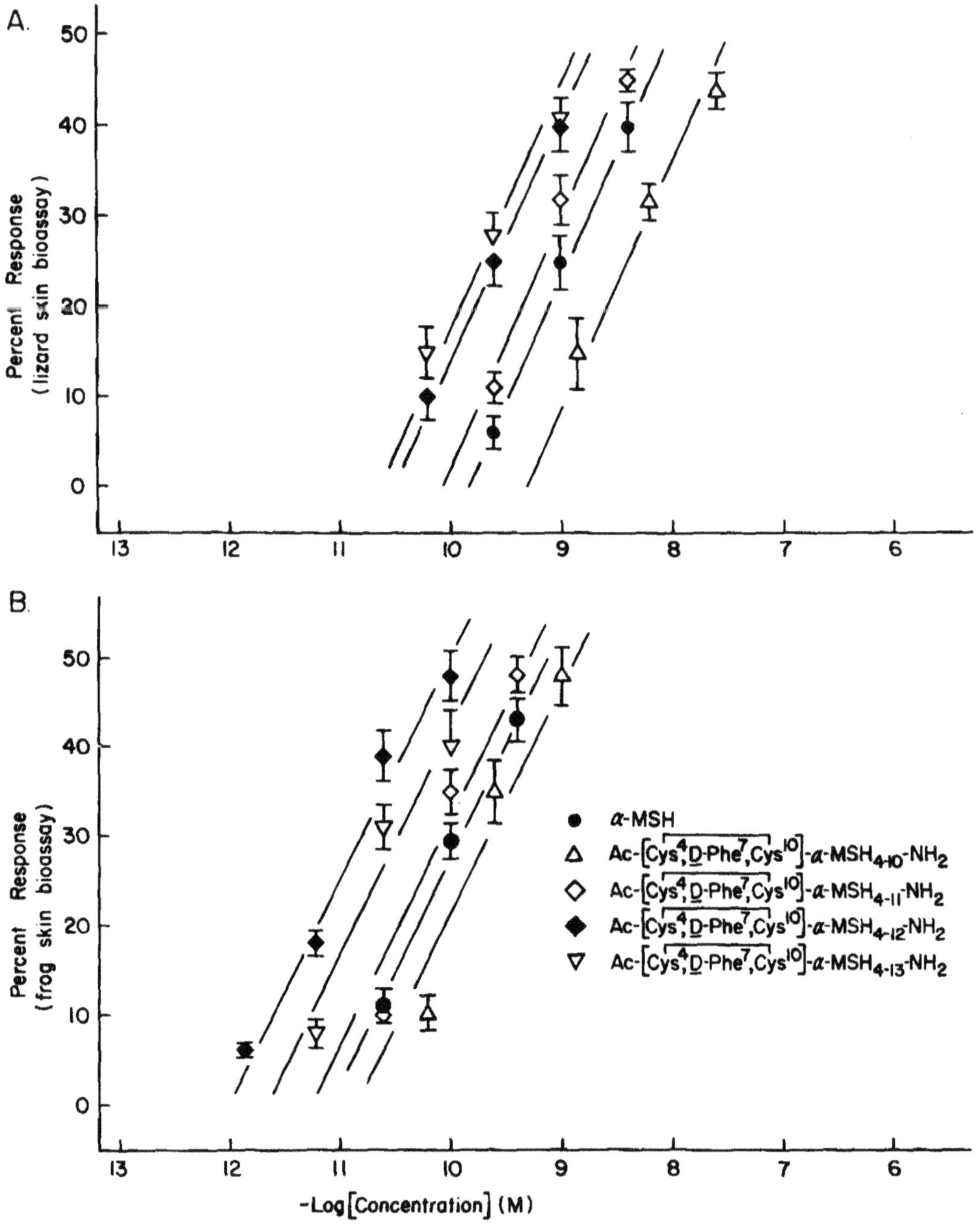

FIGURE 3. An illustration of the relative potencies of a series of cyclic cysteine D-Phe^7-substituted C-terminal fragments of α-MSH in the (A) lizard and (B) frog skin bioassays.

it lends itself to a more critical analysis of conformational structure-biological activity relationships. Various features of the cyclic molecule that may be responsible for these biological properties have been examined.

A. Substitutions for Cysteine in the Four Position

Previous studies suggested that the C-terminal tripeptide (-Ser-Tyr-Ser-) is not critical to the biological potency of cyclic melanotropins.[12,14,27,30] Thus, several substitutions for cysteine in the four position have been examined in the cyclic 4-13 analog.

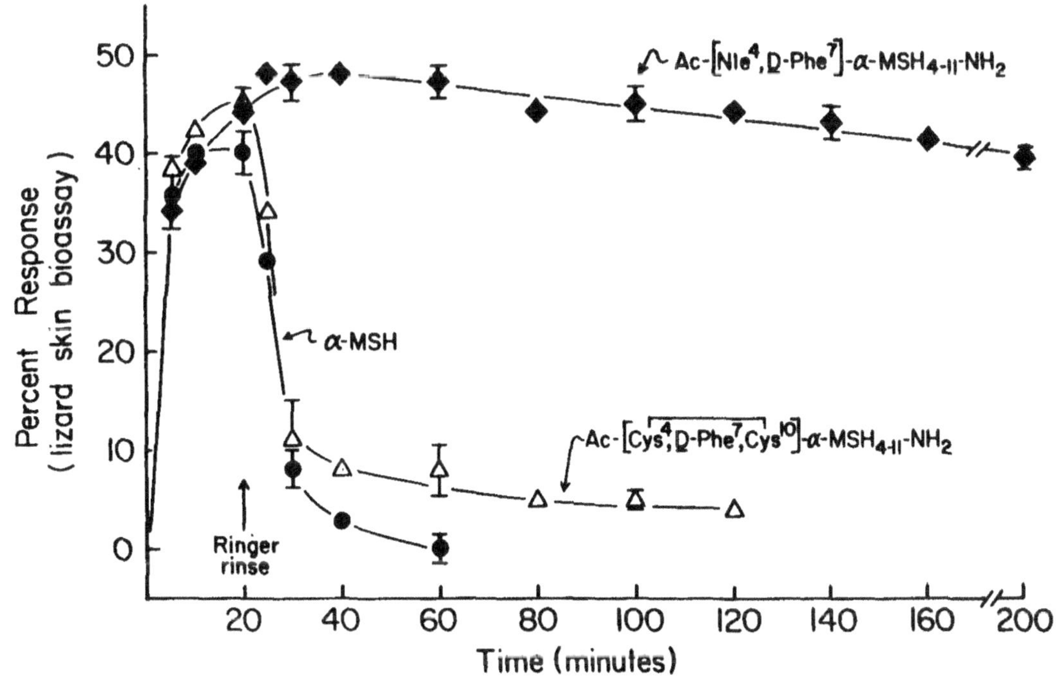

FIGURE 4. An illustration of the prolongation of biological activity of cyclic cystine D-Phe^7-substituted C-terminal fragments of α-MSH in the lizard skin bioassay, in comparison with the linear analog Ac-[Nle^4,D-Phe^7]α-MSH$_{4-11}$-NH$_2$.

Table 5
RELATIVE IN VITRO POTENCIES FOR CYCLIC DISULFIDE-LINKED MELANOTROPINS

Peptide	Biological activities[a] Frog	Biological activities[a] Lizard	Ref.
α-MSH	1.0	1.0	
Ac-[Cys^4,Cys^{10}]α-MSH$_{4-13}$-NH$_2$	30.0	0.60	12, 27, 30
Ac-[D-Cys^4,Cys^{10}]α-MSH$_{4-13}$-NH$_2$	3.0	0.70	31
[Mpa^4,Cys^{10}]α-MSH$_{4-13}$-NH$_2$	30.0	1.0	31
[Maa^4,Cys^{10}]α-MSH$_{4-13}$-NH$_2$	0.06	0.06	31
Ac-[Hcy^4,Cys^{10}]α-MSH$_{4-13}$-NH$_2$	0.06	0.70	31

[a] Biological potencies are measured relative to α-MSH (1.0) over the linear portion of the dose-response curve.

1. Ac-[D-Cys^4,Cys^{10}]α-MSH$_{4-13}$-NH$_2$

The importance of the stereochemistry of the amino acid in the four position was examined by the substitution of Cys-4 with its optical isomer D-Cys-4.[31] This modification of the parent molecule led to a tenfold loss of potency in the in vitro frog skin bioassay, but full potency was maintained on the lizard skin bioassay (Table 5). Unlike the seven position, it appears that the stereochemistry of the amino acid in the four position provides less contribution to the overall biological potency of the molecule, though such studies have not been made on linear analogs.

Further studies on the four position of the cyclic decapeptide of α-MSH (Ac-[Cys^4,Cys^{10}]α-MSH$_{4-13}$-NH$_2$) followed a "classical" strategy. The strategy utilized for the analysis of the

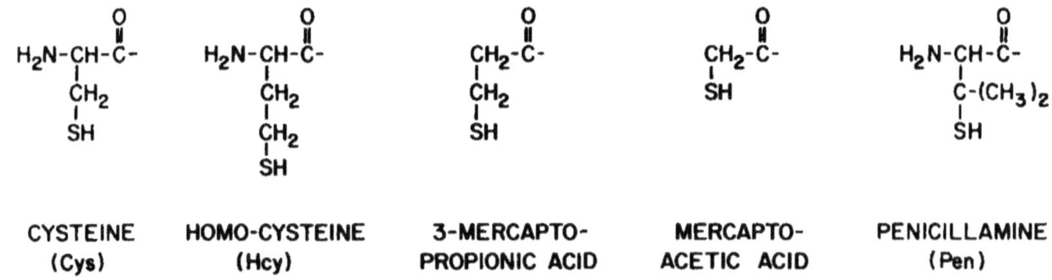

FIGURE 5. An illustration of the various substitutions for cysteine discussed in this chapter.

cyclic melanotropins paralleled the strategy of Hruby,[32] Fraefel and du Vigneaud,[33] and the strategy of Schulz and du Vigneaud[34] in their analysis of oxytocin (H-Cys-Tyr-Ile-Gln-Asn-Cys-Pro-Leu-Gly-NH_2). Unlike α-MSH the N-terminus of oxytocin is an underivatized (i.e., free amine) cysteine. Early structure-function studies of oxytocin examined the importance of subtle structural modifications of the molecule on the biological properties of the resulting analog. The N-terminal cysteine residue in Ac-[Cys^4,Cys^{10}]α-MSH_{4-13}-NH_2 provided a position for the incorporation of a series of organic compounds containing a free thiol and carboxylic acid including mercaptoacetic acid (Maa), 3-mercaptopropionic acid (Mpa), or homocysteine (Hcy) (Figure 5).[31]

2. [Mpa^4,Cys^{10}]α-MSH_{4-13}-NH_2

Early structure-activity analysis of α-MSH had shown that removal of the N-terminal acetyl group in linear analogs of α-MSH led to a tenfold reduction of in vitro frog skin potency.[12,25] However, the effect of subsequent removal of the N-terminal amine was not examined. In the cyclic 4-13 analog, replacement of cysteine with Mpa did not adversely effect the potency of the resulting analog ([Mpa^4,Cys^{10}]α-MSH_{4-13}-NH_2) in the frog skin bioassay, and, in fact, the potency was slightly enhanced in the lizard skin bioassay (Table 5).[31] These data suggest that acylation or removal of the N-terminal amine is consistent with high melanotropic potency in both biological assays.

3. [Maa^4,Cys^{10}]α-MSH_{4-13}-NH_2

The size of the intramolecular disulfide-bridged ring of Ac-[Cys^4,Cys^{10}]α-MSH_{4-13}-NH_2 (23-membered) can be reduced by the substitution of Maa in the four position for cysteine (22-membered). This substitution dramatically reduced the observed potency of the resulting analog. In the in vitro frog skin bioassay, [Maa^4,Cys^{10}]α-MSH_{4-13}-NH_2 was 500-fold less potent than Ac-[Cys^4,Cys^{10}]α-MSH_{4-13}-NH_2 and tenfold less potent than the 4-13 analog in the in vitro lizard skin bioassay (Table 5).[31] It is apparent that the magnitude of the melanotropic response is extremely sensitive to the overall three-dimensional structure of the melanotropin and the size of the ring.

4. Ac-[Hcy^4,Cys^{10}]α-MSH_{4-13}-NH_2

Likewise, enlarging the size of the intramolecular disulfide ring to a 24-membered ring by the substitution of L-cysteine with L-homocysteine (Hcy) in the four position led to a significant loss of activity on the frog skin bioassay (500-fold), but slightly enhanced the potency on the lizard skin (~1.2-fold) (Table 5).[31] The biological properties of this analog further illustrate the sensitivity of the frog receptor to the ring size of the melanotropic substrate. The above studies have shown that at the frog skin receptor the potency of cyclic melanotropins is critically related to the size of the intramolecular ring (i.e., the 23-membered ring is preferred).[31] Furthermore, the differences in biological response between the in vitro

Table 6
RELATIVE IN VITRO POTENCIES FOR CYCLIC DISULFIDE-LINKED MELANOTROPINS

Peptide	Biological activities[a]		Ref.
	Frog	Lizard	
α-MSH	1.0	1.0	
Ac-[Cys^4,Cys^{10}]α-MSH$_{4-11}$-NH$_2$	0.16	0.07	12, 30
Ac-[Pen^4,Cys^{10}]α-MSH$_{4-11}$-NH$_2$	0.5	0.03	35
Ac-[Cys^4,Pen^{10}]α-MSH$_{4-11}$-NH$_2$	0.0010	0.0017	35
Ac-[Cys^4,D-Pen^{10}]α-MSH$_{4-11}$-NH$_2$	0.0030	0.0050	35
Ac-[Pen^4,Pen^{10}]α-MSH$_{4-11}$-NH$_2$	0.0025	0.0017	35
Ac-[D-Pen^4,Pen^{10}]α-MSH$_{4-11}$.NH$_2$	0.0010	0.0014	35
Ac-[Cys^4,D-Phe^7,Cys^{10}]α-MSH$_{4-11}$-NH$_2$	2.5	3.0	30
Ac-[Pen^4,D-Phe^7,Cys^{10}]α-MSH$_{4-11}$-NH$_2$	1.0	0.10	35

[a] Biological potencies are measured relative to α-MSH (1.0) over the linear portion of the dose-response curve.

frog and lizard skin in regards to ring expansion or contraction for the cyclic derivatives of α-MSH provide additional examples revealing that these two species have melanocyte receptors with vastly differing recognition signals for melanotropins.

B. Substitution of Penicillamine for Cysteine in the Four and/or Ten Positions

An important result of the early structure-function studies with oxytocin was the substitution of penicillamine (β,β-dimethylcysteine, Pen) for cysteine in the one position and its conversion to a partial antagonist.[32] In the same regard, L- or D-penicillamine has been substitued for one or both of the cysteines in the 4-11 cyclic melanotropins. The data presented here are unpublished and preliminary; therefore, they will not be discussed extensively.[35]

Substitution of penicillamine in the four position did not affect potency of the resulting analog in either the frog or lizard skin bioassay significantly in comparison to the parent peptide (Ac-[Cys^4,Cys^{10}]α-MSH$_{4-11}$-NH$_2$). Further, substitution of D-Phe^7 for L-Phe^7 led to a significant enhancement of potency, although not as dramatic as seen in the corresponding cyclic cysteine-bridged melanotropin (Table 6).[35]

Substitution of either enantiomer of penicillamine in the ten position for cysteine led to approximately a tenfold loss in potency over the corresponding Pen^4, Cys^{10} or Cys^4, Cys^{10} analogs (Table 6). Similar potencies were observed for the bis-penicillamine-substituted analogs (Ac-[Pen^4,Pen^{10}] α-MSH$_{4-11}$-NH$_2$ and Ac-[D-Pen^4,Pen^{10}]α-MSH$_{4-11}$-NH$_2$) (Table 6). The data suggest that the penicillamine substitution in the ten position leads to a three-dimensional conformation of the melanotropin that is less compatible with high melanotropic activity.

The conformational and topological effects of the penicillamine substitutions in the four and/or ten positions are under further investigation by biophysical methods and will be reported elsewhere.

V. MISCELLANEOUS CYCLIC ANALOGS

The high potency of the cyclic 4-13 disulfide-bridged melanotropins has led to further studies of the importance of the disulfide bridge for the bioactivity of Ac-[Cys^4,Cys^{10}]α-MSH$_{4-13}$-NH$_2$. This question was initially examined by subtle modifications in the size of the intramolecular disulfide ring (see above). Further analysis involved the preparation of a

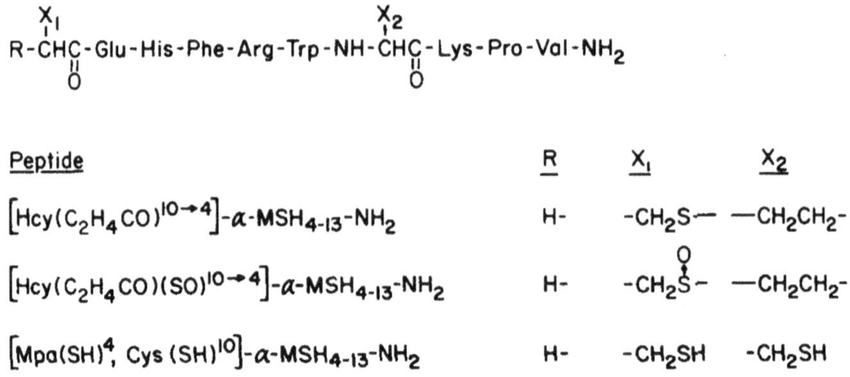

FIGURE 6. An illustration of uncommon melanotropins discussed in this chapter.

Table 7
RELATIVE IN VITRO POTENCIES FOR MISCELLANEOUS CYCLIC MELANOTROPINS

	Biological activities[a]		
Peptide	**Frog**	**Lizard**	**Ref.**
α-MSH	1.0	1.0	
Ac-[Cys^4,Cys^{10}]α-MSH_{4-13}-NH_2	30.0	0.60	27
[Mpa^4,Cys^{10}]α-MSH_{4-13}-NH_2	30.0	1.0	31
Ac-[$Hcy(C_2H_4CO)^{10\rightarrow4}$]α-$MSH_{4-13}$-$NH_2$[b]	1.0	0.20	37
Ac-[$Hcy(C_2H_4CO)(SO)^{10\rightarrow4}$]α-$MSH_{4-13}$-$NH_2$[c]	0.60	0.10	37
[$Mpa(SH)^4Cys(SH)^{10}$]α-MSH_{4-13}-NH_2[d]	0.002	0.002	37

[a] Biological potencies are measured relative to α-MSH (1.0) over the linear portion of the dose-response curve.
[b] Corresponds to the carba analog of [Mpa^4,Cys^{10}]α-MSH_{4-13}-NH_2 in which a methylene group replaces the sulfur of cystine in the ten position (see Figure 6).
[c] The oxidized analog of the carba analog described in footnote b (see Figure 6).
[b] The reduced analog of [Mpa^4,Cys^{10}]α-MSH_{4-13}-NH_2 (see Figure 6).

series of analogs with subtle modifications of the disulfide linkage by the preparation of carba analogs, oxidized carba analogs, etc. (Figure 6). This section will outline and discuss the results of these studies.

A. Carba Analogs of Ac-[Cys^4,Cys^{10}]α-MSH_{4-13}-NH_2

To further examine the importance of the sulfur atoms to high biological potency of the cyclic analogs led to the preparation of the cyclic carba 4-13 analog of [Mpa^4,Cys^{10}] α-MSH^{4-13}-NH_2. In this case the carba substitution was used to replace the sulfur of the cysteine in the ten position. This substitution was provided via a modified Hcy analog incorporated in the ten position during the solid phase synthesis utilizing a modified scheme that involved cyclization of the peptide on the resin, with subsequent cleavage from the resin and purification of the peptide.[36,37] This substitution led to a slight loss of potency in both the in vitro frog and lizard skin bioassay (~ five- to tenfold) (Table 7).

1. Oxidized Carba Analogs

The sensitivity of this system to subtle modifications of the 23-membered ring analog, [Mpa^4,Cys^{10}]α-MSH_{4-13}-NH_2, was further examined by oxidation of the sulfur in the carba analog described above. The oxidation was initiated by reaction of the purified peptide with potassium iodate in water and subsequent purification by reversed phase high-pressure liquid chromatography. This modification reduced the biological potency only slightly (~ two- to threefold) with respect to the corresponding carba analog.

B. Cyclic Analogs Chemically Reduced to the Linear Dithiol Analogs

Finally, one study examined the results of reductive oxidation of [$\overline{\text{Mpa}^4,\text{Cys}^{10}}$]α-MSH$_{4\text{-}13}$-NH$_2$ on the corresponding biological potency of the molecule. This experiment was designed to test the importance of disulfide exchange between the substrate and the melanotropin receptor to account for the high biological activity of cyclic melanotropins. The reduced analog was prepared by reduction of the parent peptide ([$\overline{\text{Mpa}^4,\text{Cys}^{10}}$]α-MSH$_{4\text{-}13}$-NH$_2$ in a solution of anhydrous liquid ammonia and sodium. The reduced peptide showed only very low potency in either the in vitro frog or lizard skin bioassay (Table 7). This low possibility residual activity could be accounted for by a small contamination of the oxidized disulfide form, although no evidence was provided for the presence of this contaminate.

VI. OTHER BIOLOGICAL ACTIVITIES OF CYCLIC MELANOTROPINS

It has been well documented that linear melanotropins can possess activity in the CNS.[3-6,38,39] (see Van der Zee et. al., Volume II, Chapter 5). In general, the potency of these analogs has been assayed in a variety of CNS assays, of which behavioral assays have been the most common.[3-6,12,38,39] Interestingly, the behavioral potencies of linear melanotropins often appears to parallel the melanotropic activity of the individual analog in both of the in vitro bioassays. For example, in the grooming assay, α-MSH itself was the most potent analog, followed closely by Ac-[Nle^4]α-MSH$_{4\text{-}13}$-NH$_2$.[38] Removal of individual amino acids from the C-terminus led to a corresponding loss of potency. The 4-10 analog was the shortest compound tested.[6,38] However, in many cases the behavioral assay results show different structure-activity relationships than those obtained in the pigmentation assays.

A careful comparison of all the C-terminal shortened cyclic melanotropins tested revealed that all cyclic analogs (e.g., 1-13, 4-13, 4-12, 4-11, and 4-10) appeared to be essentially equipotent to the native hormone α-MSH in the grooming assay.[6,38] It is possible and somewhat surprising that the two most potent analogs appear to be Ac-[$\overline{\text{Cys}^4,\text{Cys}^{10}}$]α-MSH$_{4\text{-}10}$-NH$_2$ and Ac-[$\overline{\text{Cys}^4,\text{Cys}^{10}}$]α-MSH$_{4\text{-}11}$-NH$_2$, although the data may not be statistically significant.[6,38] None of the D-phenylalanine-7 substituted analogs have been tested yet in the bahavioral assay. Further analysis of the CNS activities of cyclic as well as linear melanotropins is underway.

In a somewhat related investigation, the stability of linear and cyclic melanotropins to brain homogenates was examined.[40,41] This study was used to examine the relative stability of individual analogs to biodegradation in the in vitro system. This type of a study can be used as an initial screen to determine compounds that may be expected to be active in an in vivo system. As expected, for both linear and cyclic analogs, the substitution of D-phenylalanine in the seven position led to an enhancement of the biostability of the resulting analog.[40] These data suggest that linear or cyclic analogs of the 4-10 or 4-11 sequences with a D-phenylalanine in the seven position were essentially completely stable after 4 hr in a rat brain homogenate. These results suggest that α-melanotropins can be stabilized against biodegradation by in vivo enzymes (see Castrucci et al., Volume I, Chapter 10).

VII. SUMMARY AND FUTURE PERSPECTIVES

The high biological activities observed for cyclic $\overline{\text{Cys}^4,\text{Cys}^{10}}$ melanotropins have led to considerable research into understanding the three-dimensional structure of α-MSH responsible for these biological activities. This had been examined primarily by a classical structure-function strategy, in which subtle modifications had been made on the parent molecule and their effects on the biological properties of the resulting analogs were measured. In general, previous work has been directed at examining the conformational properties of the intramolecular disulfide ring itself, by changing its size and replacing the sulfur atoms with methylene groups. It was concluded that the size of the intramolecular bridge was the major determinant for the observed biological properties of the analogs and that any change in the size of the bridge from the original 23-membered ring led to a large loss of biological potency. The contribution to biological activity, as a result of the substitution of D-phenylalanine in the seven position, also has been examined in the cyclic series.

In general, the biological potencies of the D-phenylalanine-7-substituted $\overline{\text{Cys}^4,\text{Cys}^{10}}$ cyclic analogs paralleled those seen for the corresponding linear analogs. Very few studies have dealt with the importance of other positions or with specific conformational features which may be related to biological activity of cyclic melanotropins. Future studies will address these areas in an effort to develop more potent and receptor-specific melanotropins. These areas can be examined using individual amino acid substitutions in the 4-10 core sequence or by design of other cyclic melanotropin analogs, including side-chain to side-chain and head-to-tail cyclizations.[42]

Up to the present time, little biophysical work on the three-dimensional conformation of the cyclic melanotropins has been performed, other than by computational studies.[43,44] Since cyclization of a peptide dramatically reduces the number of conformations that are allowed, cyclic peptides (especially small ones) provide ideal analogs for a wide range of biophysical studies, including, proton NMR, X-ray diffraction, etc. Even though the highly potent cyclic melanotropins that are presently available tend to be relatively large and somewhat flexible, it is anticipated that the conformational properties of these analogs will be examined by a wide variety of biophysical techniques in order to assist with the classical structure-function analysis that has already been applied. The studies will hopefully assist in the design of smaller and more potent analogs, which will be more suitable for rigorous biophysical analysis.

ACKNOWLEDGMENTS

This work was supported by U.S. Public Health Service Grant AM 17420 (VJH) and by grants from the National Science Foundation (MEH and VJH). We thank Karen Burgan for typing and helping to edit this manuscript.

REFERENCES

1. **Hadley, M. E. and Bagnara, J. T.,** Regulation of release and mechanism of action of MSH, *Am. Zool.*, 15 (Suppl. 1), 81, 1975.
2. **Hadley, M. E., Heward, C. B., Hruby, V. J., Sawyer, T. K., and Yang, Y. C. S.,** Biological actions of melanocyte stimulating hormone, in *Peptides of the Pars Intermedia*, Evered, D. and Lawrenson, G., Eds., Pitman Medical, London, 1981, 244.
3. **LaHoste, G. J., Olson, G. A., Kastin, A. J., and Olson, R. D.,** Behavioral effects of melanocyte stimulating hormone, *Neurosci. Biobehav. Rev.*, 4, 9, 1980.
4. **O'Donohue, T. L. and Dorsa, D. M.,** The opiomelanotropinergic neuronal and endocrine systems, *Peptides*, 3, 353, 1982.
5. **O'Donohue, T. L., Handelmann, G. E., and Chaconas, T., Miller, R. L., and Jacobowitz, D. M.,** Evidence that N-acetylation regulates the behavioral activity of α-MSH in the rat and human central nervous system, *Peptides*, 2, 333, 1981.

6. **Sawyer, T. K., Cody, W. L., Knittel, J. J., Hruby, V. J., Hadley, M. E., Hirsch, M. D., and O'Donohue, T. L.,** Design of conformationally-restricted cyclic α-melanotropins: comparison of melanocyte stimulating and behavioral activities, in *Peptides: Structure and Function, Proc. 8th Am. Peptide Symp.*, Hruby, V. J. and Rich, D. H., Eds., Pierce Chemical Co., Rockford, Ill. 1983, 323.
7. **Wright, M. R. and Lerner, A. B.,** On the movement of pigment granules in frog melanocytes, *Endocrinology*, 66, 599, 1960.
8. **Bregman, M. D., Sawyer, T. K., Hadley, M. E., and Hruby, V. J.,** Adenosine and divalent cation effects on S-91 melanoma adenylate cyclase, *Arch. Biochem. Biophys.*, 200, 1, 1980.
9. **Fuller, B. B. and Hadley, M. E.,** Transcriptional and translational requirements for MSH stimulation of tyrosinase in melanoma cells, *Pigment Cell*, 4, 97, 1979.
10. **Fuller, B. B. and Viskochil, D. H.,** The role of RNA and protein synthesis in mediating the action of MSH on mouse melanoma cells, *Life Sci.*, 24, 2405, 1979.
11. **Sawyer, T. K., Sanfilippo, P. J., Hruby, V. J., Engel, M. H., Heward, C. B., Burnett, J. B., and Hadley, M. E.,** 4-Norleucine,7-D-phenylalanine-α-melanocyte-stimulating hormone: a highly potent α-melanotropin with ultra long biological activity, *Proc. Natl. Acad. Sci. U.S.A.*, 77, 5754, 1980.
12. **Hruby, V. J., Wilkes, B. C., Cody, W. L., Sawyer, T. K., and Hadley, M. E.,** Melanotropins: structural, conformational and biological considerations, in *Peptide and Protein Reviews*, Vol. 3, Hearn, M. T. W., Ed., Marcel Dekker, New York, 1984, 1.
13. **Wilkes, B. C., Sawyer, T. K., Hruby, V. J., and Hadley, M. E.,** Comparative biological activities of potent active-site analogs of α-melanotropin, *Intl. J. Peptide Protein Res.*, 23, 621, 1984.
14. **Cody, W. L., Mahoney, M., Knittel, J. J., Hruby, V. J., Castrucci, A. M., de L., and Hadley, M. E.,** Cyclic melanotropins. IX. D-phenylalanine[7] analogues of the active site sequence, *J. Med. Chem.*, 28, 583, 1984.
15. **Bool, A. M., Gray, G. H., II, Hadley, M. E., Heward, C. B., Hruby, V. J., Sawyer, T. K., and Yang, Y. C. S.,** Racemization effects on melanocyte-stimulating hormones and related peptides, *J. Endocrinol.*, 88, 57, 1981.
16. **Engel, M. H., Sawyer, T. K., Hadley, M. E., and Hruby, V. J.,** Quantitative determination of amino acid racemization in heat-alkali treated melanotropins: implications for peptide hormone structure-function studies, *Anal. Biochem.*, 166, 303, 1981.
17. **Hruby, V. J., Sawyer, T. K., Yang, Y. C. S., Bregman, M. D., Hadley, M. E., and Heward, C. B.,** Synthesis and structure-function studies of melanocyte-stimulating hormone analogues modified in the 2 and 4(7) positions: comparison of activities on frog skin melanophores and melanoma adenylate cyclase, *J. Med. Chem.*, 23, 1432, 1980.
18. **Sawyer, T. K., Darman, P. S., Hruby, V. J., and Hadley, M. E.,** Conformation-activity studies of α-melanotropin exhibiting high potency and full biological activity, in *Peptides: Synthesis-Structure-Function, Proc. 7th Am. Peptide Symp.*, Rich, D. H. and Gross, E., Eds., Pierce Chemical, Rockford, Ill., 1981, 387.
19. **Sawyer, T. K., Darman, P. S., Hruby, V. J., and Hadley, M. E.,** [4-Half-cystine, 10-half-cystine]-melanocyte stimulating hormone: a cyclic α-melanotropin exhibiting superagonistic biological activity, *Proc. Natl. Acad. Sci. U.S.A.*, 79, 1751, 1982.
20. **Hruby, V. J.,** Conformational and dynamic considerations for the design of peptide superagonists and antagonists, in *Conformationally Directed Drug Design*, ACS Symp. Ser. No. 251, Vida, J. A. and Gordon, M., Eds., Washington, D. C., 1984, 9.
21. **Hruby, V. J. and Hadley, M. E.,** Binding and information transfer in conformationally restricted peptides, in *Design and Synthesis of Organic Molecules Based on Molecular Recognition. Proc. 18th Solvay Conf. on Chemistry*, van Binst, G., Eds., Springer-Verlag, Heidelberg, 1986, 269.
22. **Castrucci, A. M., de L., Hadley, M. E., and Hruby, V. J.,** Melanotropin bioassays: *in vitro* and *in vivo* comparisons, *Gen. Comp. Endocrinol.*, 55, 104, 1984.
23. **Eberle, A. and Schwyzer, R.,** Hormone-receptor interaction. demonstration of two message sequences (active sites) in α-melanotropin, *Helv. Chim. Acta*, 58, 1528, 1975.
24. **Wilkes, B. C., Hruby, V. J., and Hadley, M. E.,** Synthesis and structure-function activity studies of highly potent α-MSH analogues, presented at the 8th Am. Peptide Symp., Tucson, May 22 to 27, 1983, 139.
25. **Hofmann, K. and Yajima, H.,** Synthetic pituitary hormones, *Recent Progs. Horm. Res.*, 18, 41, 1962.
26. **Schwyzer, R., Costopanagiotis, A., and Sieger, P.,** Zwei Synthesen des α-Melanotropins (α-MSH) mit Hilfe leicht entfernbarer Schutzgruppen, *Helv. Chim. Acta*, 46, 870, 1963.
27. **Knittel, J. J., Sawyer, T. K., Hruby, V. J., and Hadley, M. E.,** Structure-activity studies of highly potent cyclic [Cys[4],Cys[10]] melanotropin analogues, *J. Med. Chem.*, 28, 125, 1983.
28. **Sawyer, T. K., Hruby, V. J., Wilkes, B. C., Draelos, M. T., Hadley, M. E., and Bergsneider, M.,** Comparative biological activities of highly potent active-site analogues of α-melanotropin, *J. Med. Chem.*, 25, 1022, 1982.

29. **Wilkes, B. C., Sawyer, T. K., Hruby, V. J., and Hadley, M. E.,** Comparative biological activities of potent active-site analogues of α-melanotropin. Effect of tyrosine substitution at position-4, *Intl. J. Peptide Protein Res.*, 23, 621, 1984.
30. **Cody, W. L., Wilkes, B. C., Muska, B. J., Hruby, V. J., de L. Castrucci, A. M., and Hadley, M. E.,** Cyclic melanotropins. V. Importance of the C-terminal tripeptide (Lys-Pro-Val), *J. Med. Chem.*, 27, 1186, 1984.
31. **Lebl, M., Cody, W. L., Wilkes, B. C., Hruby, V. J., de L., Castrucci, A. M., and Hadley, M. E.,** Cyclic melanotropins, VII. Modified ring structures, synthesis and biological activity, *Intl. J. Peptide Protein Res.*, 24, 472, 1984.
32. **Hruby, V. J.,** Structures of the neurohypophysical hormones and analogues, and implications for hormone-receptor interactions, in *Biochemical Actions of Hormones*, Vol. 13, Litwack, G., Ed., Academic Press, New York, 1986, 191.
33. **Fraefel, W. and du Vigneaud, V.,** [1-(Mercaptoundecanoic acid)]-oxytocin a 28-membered ring homolog of deamino-oxytocin, *J. Am. Chem. Soc.*, 92, 4426, 1970.
34. **Schulz, H. and du Vigneaud, V.,** Synthesis of 1-*L*-penicillamine-oxytocin, 1-D-penicillamine-oxytocin, and 1-deaminopenicillamine-oxytocin, potent inhibitors of the oxytocic response of oxytocin, *J. Med. Chem.*, 9, 647, 1966.
35. **Cody, W. L.,** Synthesis, Biological Activity and Conformational Analysis of Fragment Analogues of Alpha-Melanotropin, Ph.D. thesis, University of Arizona, Tucson, 1985.
36. **Lebl, M. and Hruby, V. J.,** Synthesis of cyclic peptides by solid phase methodlogy, *Tetrahedron Lett.*, 25, 2067, 1984.
37. **Lebl, M., Cody, W. L., Wilkes, B. C., Hruby, V. J., de L., Castrucci, A. M., and Hadley, M. E.,** Cyclic melanotropins. VIII. Modifications of the disulfide bridge in cyclic melanotropins, *Collect. Czech. Chem. Commun.*, 49, 2680, 1984.
38. **Hirsch, M. D., O'Donohue, T. L., Wilson, R., Sawyer, T. K., Hruby, V. J., Hadley, M. E., Cody, W. L., Knittel, J. J., and Crawley, J. N.,** Structural and conformational modifications of α-MSH/ $ACTH_{4-10}$ provide melanotropin analogues with highly potent behavioral activities, *Peptides*, 5, 1197, 1984.
39. **Van Nispen, J. W. and Greven, H. M.,** Structure-behavioral activity relationships of peptides derived from ACTH: some stereochemical considerations, in *Conformationally Directed Drug Design*, ACS Symp. Ser. 251, Vida, J. A. and Gordon, M. Eds., Washington, D.C., 1984, 153 and references therein.
40. **Akiyama, K., Yamamura, H. I., Wilkes, B. C., Cody, W. L., Hruby, V. J., de L. Castrucci, A. M., and Hadley, M. E.,** Relative stability of α-melanotropin and related analogues to rat brain homogenates, *Peptides*, 5, 1191, 1984.
41. **de L. Castrucci, A. M., Hadley, M. E., Sawyer, T. K., and Hruby, V. J.,** Enzymological studies of melanotropins, *Comp. Biochem. Physiol.*, 78B, 519, 1984.
42. **Hruby, V. J.,** Conformational restrictions of biologically active peptides via amino acid side chain groups, *Life Sci.*, 31, 189, 1982.
43. **Nikiforovich, G., Rozenblit, I., Liepina, I., and Chipens, G.,** Conformational calculations and biologically active conformatios of cyclopeptides, in *Peptides 1982, Proc. 17th Eur. Peptide Symp.*, Blaha, K. and Malon, P., Eds., Walter de Gruyter, Berlin, 1983, 735.
44. **Nikiforovich, G. V., Rozenblit, M. D., Shenderovich, M. D., and Chipens, G. I.,** Possible bioactive conformations of α-melanotropins, *FEBS Lett.*, 170, 315, 1984.

Chapter 7

MELANOTROPIN THREE-DIMENSIONAL STRUCTURAL STUDIES BY PHYSICAL METHODS AND COMPUTER-ASSISTED MOLECULAR MODELING

Wayne L. Cody, J. W. Sam Stevenson, Fahad Al-Obeidi, Elizabeth E. Sugg, and Victor J. Hruby

TABLE OF CONTENTS

I. Introduction 94
- A. History 94
  - 1. Biological Significance of α-Melanocyte-Stimulating Hormone (α-MSH) 94
  - 2. Early Structure-Function Studies 94
  - 3. Summary of Biophysical Methods Applied to the Analysis of α-MSH 94

II. Biophysical Analysis of Analogs of α-MSH 94
- A. Reversed Phase High Performance Liquid Chromatography Studies (RP-HPLC) 94
  - 1. Application of RP-HPLC to the Analysis of α-MSH Conformation 94
    - a. Linear Analogs of α-MSH 95
    - b. Cyclic Analogs of α-MSH 96
- B. Circular Dichroism (CD) Studies 96
  - 1. Application of CD to the Analysis of α-MSH Conformation 96
- C. Computer-Assisted Molecular Modeling Studies — Theoretical Studies of α-MSH and its Analogs 96
- D. Nuclear Magnetic Resonance (NMR) Studies 100
  - 1. Application of NMR to the Study of Peptide Conformation 100
  - 2. Application of NMR to the Analysis of Analogs of α-MSH 100
    - a. Early Carbon 13 Studies 101
    - b. Proton NMR Analysis of ACTH Analogs Related to α-MSH 101
    - c. NMR Analysis of Linear 4-11 Analogs of α-MSH 103

III. Summary and Future Perspectives 106

References 107

I. INTRODUCTION

A. History

1. Biological Significance of α-Melanocyte-Stimulating Hormone (α-MSH)

α-MSH or α-melanotropin, (Ac-Ser^1-Tyr^2-Ser^3-Met^4-Glu^5-His^6-Phe^7-Arg^8-Trp^9-Gly^{10}-Lys^{11}-Pro^{12}-Val^{13}-NH_2), is synthesized in the vertebrate pars intermedia and in the brain.[1,2] Biologically, α-MSH is primarily known for its ability to stimulate integumental melanocytes, but it has also been implicated in a variety of other physiological and neurological processes.[3-6]

2. Early Structure-Function Studies

Until recently, attempts to understand the preferred three-dimensional structure of α-MSH and its relationship to biological activity were based on the substitution or deletion of individual amino acid residues in α-MSH and/or shorter linear fragments.[7,8] Structure-activity relationships in these linear analogs showed the importance of stereochemistry of the amino acid in position seven to biological potency and prolongation of the melanotropic effect,[7,9-12] and it was suggested that this might be related to specific conformational effects.[7,9,13] Since linear peptides are inherently flexible, interpretation of conformational data on the linear D-Phe^7 analogs is difficult.

There have been no significant advances in the determination of the biologically active conformation of linear α-MSH analogs until recently. However, the preparation of a series of highly potent, cyclic, conformationally restricted analogs, cyclized by the pseudoisosteric substitution of cysteine in the four and ten positions for Met^4 and Gly^{10},[11,12,14-16] was based on conformational considerations including (1) the high potency of the D-Phe^7-containing series (D-amino acids can stabilize a reverse turn)[17] and (2) Chou-Fassman calculations and model building.[14,18] The cyclic compounds generally possessed significantly greater biological potency than their pseudoisosteric linear counterparts both in vitro and in vivo.[7,14-16] Since cyclization leads to considerble restriction of the conformational flexibility of the peptide backbone and, to a lesser degree, the side-chain groups,[19] cyclic peptides lend themselves to a more complete evaluation of their biologically active three-dimensional topologies. Even though the size of the intramolecular ring in a α-MSH analogs is relatively large (23 atoms), extensive biophysical examination may provide critical clues to the preferred biologically active conformation.

3. Summary of Biophysical Methods Applied to the Analysis of α-MSH

The relationship between the three-dimensional conformation and biological activity in α-MSH has been examined by a variety of researchers using numerous biophysical methods including (1) reversed phase high-performance liquid chromatography (RP-HPLC),[20,21] (2) circular dichroism (CD),[22] (3) computer-assisted molecular modeling,[23-25] and (4) proton[26-29] and carbon-13 nuclear magnetic resonance (NMR) spectroscopy. The purpose of this chapter will be to discuss the results obtained to date, critically analyze the results, and provide stimulus for future research.

II. BIOPHYSICAL ANALYSIS OF ANALOGS OF α-MSH

B. Reversed Phase High Performance Liquid Chromatography Studies (RP-HPLC)

1. Application of RP-HPLC to the analysis of α-MSH Conformation

Reversed phase high performance liquid chromatography (RP-HPLC) has found significant application in the purification of biologically active compounds, such as peptides and proteins. This technique has been utilized for the separation of biologically active peptides which differ only slightly in their three-dimensional structure or overall lipophilicity.[20,21,31]

Table 1
RETENTION TIMES EXPRESSED AS K′ VALUES FOR SELECTED 4-11 FRAGMENTS OF Ac-[Nle^4]α-MSH$_{4\text{-}11}$-NH$_2$ WITH ENANTIOMERIC SUBSTITUTIONS IN THE SEVEN POSITION OF Ac-Nle^4-Glu^5-His^6-R^7-Arg^8-Trp^9-Gly^{10}-Lys^{11}-NH$_2$ IN VARYING ISOCRATIC MOBILE PHASES

Compound[a]	Conditions[b]				
R	A	B	C	D	E
Gly	1.38	1.15	0.59	1.53	1.52
D-Ala	1.61	1.30	0.82	1.53	1.94
Ala	1.77	1.30	0.91	1.67	2.12
D-Tyr	2.69	1.60	1.82	1.80	3.47
Tyr	4.15	1.75	4.00	2.67	5.29
D-Pgl[c]	3.62	1.85	2.54	2.07	4.00
Pgl	7.69	2.30	5.82	2.93	9.24
D-Phe	8.85	2.55	7.91	3.40	12.41
Phe	15.77	3.25	15.18	5.07	25.24
D-Pnp[c]	N.D.[d]	4.45	N.D.[d]	4.53	N.D.[d]
Pnp	16.54	3.50	N.D.[d]	4.40	20.40

[a] All compounds were chromatographed on a C18 Vydac® (25.0 × 0.46 cm I.D.) 16 μm reversed phase column (#830822).
[b] A = 0.1% aqueous trifluoroacetic acid-acetonitrile (80:20); B = 0.1% aqueous hexafluorobutyric acid-acetonitrile (70:30); C = 0.25 *M* aqueous triethylammonium phosphate (pH 2.2)-acetonitrile (86:14); D = 0.1% aqueous hexafluorobutyric acid-methanol (50:50); E = 0.1% aqueous trifluoroacetic acid-methanol (68:32).
[c] Pgl = phenylglycine, Pnp = para-nitrophenylalanine.
[d] N.D. = not determined.

From Cody, W. L., Wilkes, B. C., and Hruby, V. J., *J. Chromatogr.*, 314, 313, 1984. With permission.

Often, a subtle structural or conformational modification of a peptide will have unpredictable results on the observed RP-HPLC retention time. However, careful examination of the chromatographic properties of a series of peptides with minimal modifications can provide some conformational information. For example, if replacement of a particular amino acid residue or functional group by a more lipophilic moiety leads to increased retention times on RP-HPLC, then the conformations of those peptides are probably related. However, if the substitution of a more lipophilic moiety results in a decreased retention time, then it can be inferred that the conformations have changed.[31] Changing the stereochemistry of an individual amino acid in a peptide would not change the lipophilicity of that residue, but could influence the overall lipophilicity of a peptide if a conformational change is indicated by this modification.[20,21,31]

a. Linear Analogs of α-MSH

The importance of the seven position on the conformation of norleucine^4 (Nle^4)-substituted 4-11 linear analogs substitued with a variety of amino acids of both the D- and L-configurations in the seven position (Table 1),[21] was examined on a Vydac C18® reversed phase column utilizing five different mobile phases (see Table 1 for the specific conditions and compounds examined). In general, analogs of α-MSH with a D-amino acid in the seven position eluted earlier than those containing an L-amino acid. The single exception was the L- or D-*p*-nitrophenylalanine substitution in the seven position. The data suggest that in the five solvent systems the three-dimensional structures of the fragment analogs depend on the stereochemistry of the amino acid in the seven position. It appears that the substitution of a D-amino

ACTH(6-10)	His-Phe-Arg-Trp-Gly
[D-His^{6}]ACTH(6-10)	D-His-Phe-Arg-Trp-Gly
[D-Phe^{7}]ACTH(6-10)	His-D-Phe-Arg-Trp-Gly
[D-Arg^{8}]ACTH(6-10)	His-Phe-D-Arg-Trp-Gly
[D-Trp^{9}]ACTH(6-10)	His-Phe-Arg-D-Trp-Gly
[D-His^{6}, D-Phe^{7}, D-Arg^{8}, D-Trp]ACTH(6-10)	D-His-D-Phe-D-Arg-D-Trp-Gly

FIGURE 1. Analogs of ACTH(6-10) examined by both (CD)[27] and proton (^{1}H-NMR) spectroscopy.[28,29]

acid in the seven position of linear analogs generally leads to a conformation which is less lipophilic than that when an L-amino acid is present.

b. Cyclic Analogs of α-MSH

A similar study examined the effects of the stereochemistry of the amino acid in the four and seven positions of cyclic disulfide-linked fragments of α-MSH of varying length (i.e., 4-10, 4-11, 4-12, and 4-13) on their chromatographic behavior in a variety of solvent systems on two C18 RP-HPLC columns (Vydac® [16 μm] or Altech® [5 μm])[20] (data not shown). Similar to the linear analogs, incorporation of a D-amino acid (Phe) in the seven position of the cyclic analogs led to a less lipophilic peptide than the corresponding L-amino-acid-containing analog. Likewise, the D-Cys^{4} substitution for L-Cys^{4} led to a loss of retention on RP-HPLC, suggesting both a change in overall conformation and a corresponding decrease in lipophilicity.

B. Circular Dichroism (CD) Studies

1. Application of CD to the Analysis of α-MSH Conformation

CD has proven to be a powerful tool in the analysis of the conformational properties of peptides or proteins.[32] However, conformational information about individual amino acids is often difficult to obtain. In general, this method has been applied to an analysis of various aspects of secondary structure for large peptides or proteins. In many applications, CD has provided critical information regarding the three-dimensional properties of a given protein.[32]

The ability of CD to provide critical insight into the conformations of small linear peptides is still debated. Nevertheless, pentapeptide fragments of adrenocorticotropin (ACTH) (6-10) (H-His^{6}-Phe^{7}-Arg^{8}-Trp^{a}-Gly^{10})[22,28,29] have been examined by CD with D-amino acids substituted for each amino acid in this sequence (Figure 1).[22]

The CD spectra of six analogs of ACTH(6-10) (Figure 1) were examined in water at room temperature.[22] The melanotropic activities of the analogs covered a large potency range. In all cases, it was shown that the pentapeptides exhibited no apparent secondary structure; thus, these studies provided no insight into the conformation-activity relationships. These observations may be a result of the flexibility of these linear peptides or may be a function of the particular solvent system used for the study.

C. Computer-Assisted Molecular Modeling Studies — Theoretical Studies of α-MSH and its Analogs

The molecular basis of recognition of peptide hormones and neurotransmitters is undoubtedly related to their three-dimensional structure. However, the high flexibility of these peptide hormones and the dependency of their conformation on environment creates a major problem in directly transferring X-ray, theoretical, or experimental studies to possible conformations at the receptor level.

Theoretical calculations of peptide conformations have been pioneered by Scheraga,[33] Ramachandran and co-workers[34] and Ramachandran and Sasisekharan.[35] These calculations have been used to provide information when conformational information has not been available by experimental techniques or when it is needed to interpret experimental results.[36]

Modern computer technology has made it possible to extend theoretical conformational studies of peptides to examine their molecular dynamics and internal energetics,[37,38] and energy minimization of peptide structure has been used in several laboratories to correlate the low-energy structures with the biologically active conformation.[39,40]

A few studies of α-MSH and its linear and cyclic analogs have been carried out in an effort to understand the bioactive conformation of this peptide hormone. Rigid-geometry-minimization techniques have been used throughout these studies.[23-25]

Calculations started with α-MSH having various typical, standard secondary structures for the peptide backbone, including fully extended, α-helical, type II β-turns, and C_7-helical conformations. The side-chains were placed in several possible conformations for each of the backbone conformations. These side-chain conformations were chosen from standard sets of chi (χ) angles equal to +60, −60, and 180° (i.e., gauche [+], gauche [−], and trans). A number of different side-chain conformations were used as initial structures according to the nature of the final backbone conformational angles. The initial calculations were carried out on fragments 1-4, 1-6, 5-13, and 12-13 of α-MSH. In the case of cyclic analogs, a ring-closing potential was added. Any conformation with $\Delta E < 10$ kcal/mol was retained. This approach has certain disadvantages. First, it is important to note that, in the course of using peptide fragments to build up the full α-MSH structure, it is necessary to retain many low-energy conformations in the allowed sets of fragments for building up the longer structures, since premature discarding of some of these structures could lead to erroneous results at later stages of the calculations. Moreover, at the tetrapeptide or pentapeptide stage, several quite different conformations are likely to have very similar energies. Therefore, the "right" conformation for these small oligopeptides cannot be properly selected as the appropriate low-energy one until later stages of the computations when the full peptide of interest is being examined and its conformations can be influenced by long-range as well as short- and medium-range interactions.

The theoretically generated structures of α-MSH[23-25,41] placed the Glu^5 and Arg^8 side-chains close to each other in most of the energy minimized structures, while the Glu^5 and Lys^{11} side-chains were further apart in most of the structures that had been suggested previously[13,15,18] using other approaches. Unfortunately, the effect of the D-Phe^7 substitution on these low-energy conformations of α-melanotropins was not examined.

In the case of Ac-[$\overline{Cys^4,Cys^{10}}$]α-MSH analogs (1-13, 4-13, and 4-10), a β-turn was observed for the His^6-Phe^7-Arg^8-Trp^9 sequence (Table 2). These results are consistent with the NMR studies of several fragments of α-MSH. Most of the conformationally allowed structures showed a close proximity of the side-chains of Glu^5 and Arg^8, but some of the calculated structures for either α-MSH or its cyclic analogs indicated a proximity of the Glu^5 and Lys^{11} side-chain groups.

The theoretical calculations indicated that a β-turn at the active core of α-MSH (-His^6-Phe^7-Arg^8-Trp^9-) is one of the possible conformationally allowed structures, but that several others also were possible. A major weakness of these calculations was the lack of information regarding the conformation of the amino acid side-chain moieties. For example, these calculations did not show the relative orientations of the side-chain groups or even the orientation of the Glu^5 and Lys^{11} side-chain group in the cases where the Glu^5 and Arg^8 ionic side-chain groups were predicted to be close to each other. Thus, the overall topology of the calculated structures cannot be deduced from these studies. Nonetheless, these computational

Table 2
CALCULATED CONFORMATIONS OF α-MELANOTROPINS

Compound	Residue													Distance (Å)		
	Ser[1]	Tyr[2]	Ser[3]	Met[4](Cys)	Glu[5]	His[6]	Phe[7]	Arg[8]	Trp[9]	Gly[10](Cys)	Lys[11]	Pro[12]	Val[13]	ΔE (kcal/mol)	C^{δ}Glu[5]-C^{η}Arg[8]	C^{δ}Glu[5]-N^{ϵ}Lys[11]
α-MSH	B[a]	L	R	R	B	B	R	R	B	L	B	R	B	0.0	3.3	8.8
Ac-[Cys[4],Cys[10]]α-MSH_{1-13}-NH_2	B	B	B	R	B	R	R	B	R	B	B	R	B	0.0	4.8	3.6
Ac-[Cys[4],Cys[10]]α-MSH_{4-13}-NH_2				R	L	L	L	R	R	B	B	R	B	0.0	3.6	9.9
				R	B	R	R	B	R	B	B	R	B	1.9	4.7	3.6
				B[b]	B	R	R	R	B	B	B	B	B	7.0	4.0	18.1
Ac-[Cys[4],Cys[10]]α-MSH_{4-10}-NH_2				B	B	L	L	R	B	B				0.0	4.5	—
Ac-α-MSH_{6-9}-NH_2						B	R	R	B				0.0			

[a] The peptide backbone angles were defined as follows: B ($\theta \sim -140°$, $\psi \sim 140°$); R ($\theta \sim -60°$, $\psi \sim -60°$); L ($\theta \sim 60°$, $\psi \sim 60°$); and H ($\theta \sim -80°$, $\psi \sim -80°$) (for the Gly[10] residue). All side-chain conformations were defined to be $\chi 1$ + 60°, 180°, and −60°.

[b] These structures are shown in a stereo view in Figure 2.

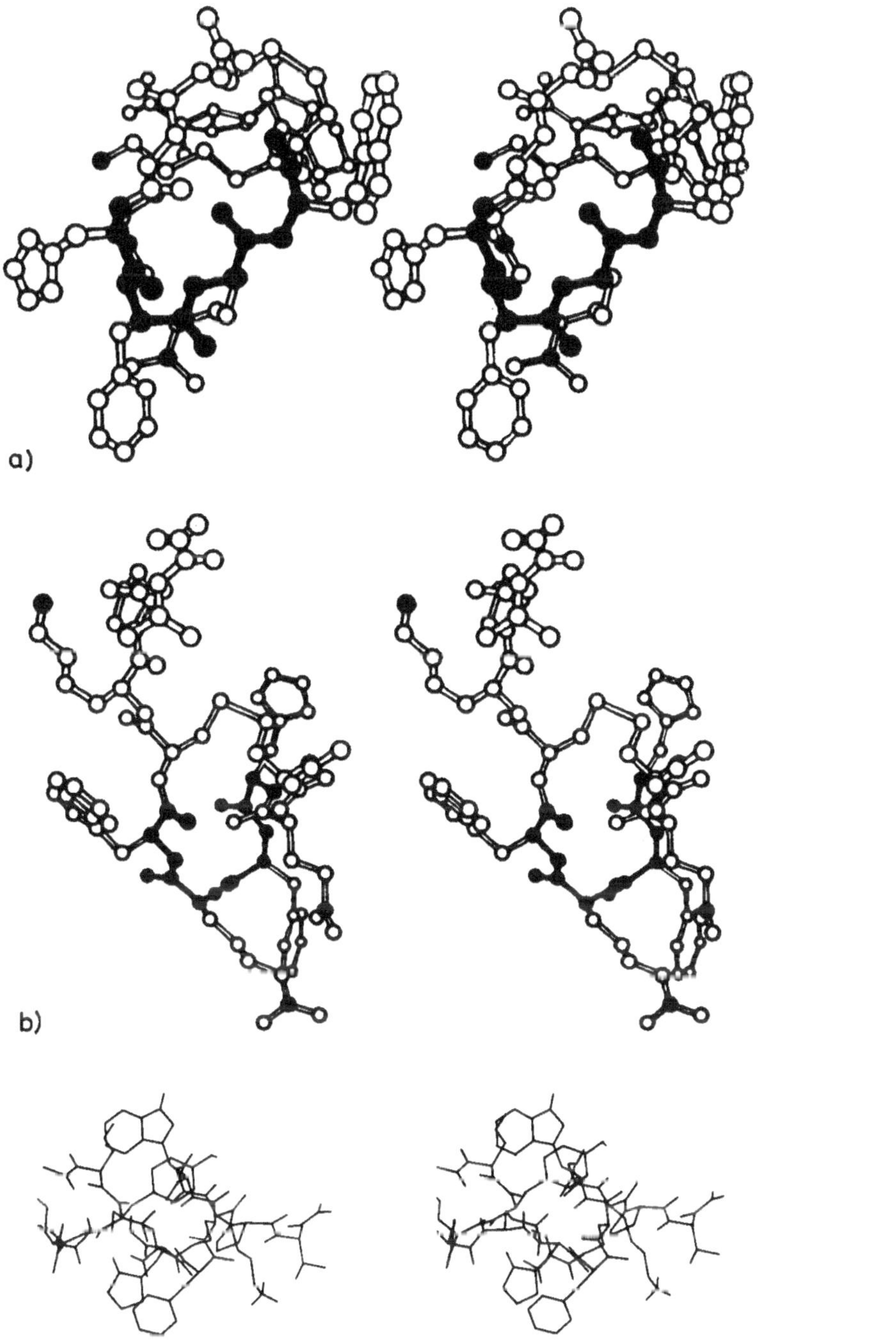

FIGURE 2. Stereo views of hypothetical bioactive conformations of Ac-[$\overline{Cys^4,Cys^{10}}$]α-$MSH_{4-13}$-$NH_2$ (a, b — top) according to Reference 25, and of α-MSH according to Reference 23 (bottom).

predictions of α-MSH and its analogs have provided a starting point for further investigations of α-MSH conformations, and in some cases appear to be consistent with some of the physical studies on α-MSH structure. The structures in Figure 2 and Table 2 summarize the results from the various theoretical studies. Dynamic simulations and graphic analysis of α-MSH and its analogs are in progress in our laboratory at the University of Arizona in an effort to further examine possible α-melanotropin conformations.

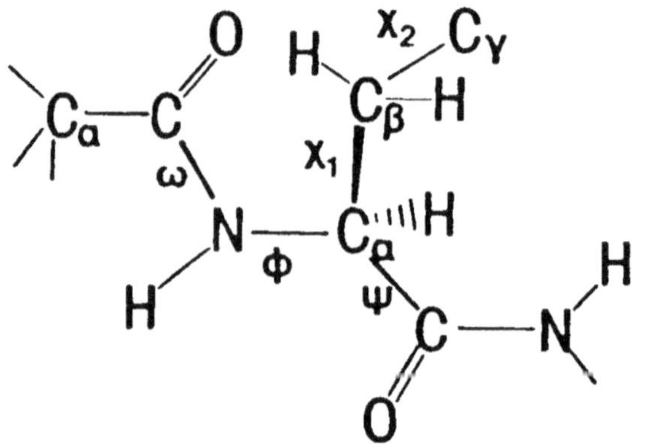

FIGURE 3. Illustration of angles which define peptide conformation (ϕ, ψ, ω, and χ).

D. Nuclear Magnetic Resonance Studies

1. Application of NMR to the Study of Peptide Conformation

NMR spectroscopy has evolved to a state which makes it an excellent method for determining the solution conformation of peptides.[42,43] However, determination of the conformation of peptides in solution is still a difficult task, especially for linear peptides which usually possess a myriad of conformational states. Nonetheless, in favorable cases, important conformational information can be obtained even for linear peptides.

Amino acid and peptide conformations are defined by the torsional angles Φ Ψ and ω which describe the rotation about the backbone bonds, N-C^{α}, C^{α}-C' and C'-N, respectively (See Figure 3). Additionally, the torsional angles, χ^1, χ^2, etc., which describe the rotation about the side-chain bonds, provide additional conformational information.

Given the many possible conformations, distinct conformations can generally be observed by NMR only when the energy barrier for interconversion is on the order of 20 kJ/mol or greater. The free enthalpy of rotation about the resonance-hybridized amide bond is on the order of 75 to 85 kJ/mol,[44] and hence this bond (the ω angle) is usually (except for Pro) found in the trans conformation. In addition, the available conformational space of the peptide backbone and the side-chain residues may be reduced by (1) intramolecular hydrogen bonding, (2) steric factors which may prohibit certain intramolecular interactions, and (3) appropriate arrangement of the molecular and chemical environment so as to promote important intramolecular interactions.

A general strategy for the eludication of a peptide's solution conformation using NMR techniques must first involve a reliable and unambiguous assignment of the signals[42] and determination of the scalar coupling constants (J) which can be used to calculate the bond torsional angles via Karplus-like equations.[45,46] Another valuable tool for determining the intramolecular spatial proximity of groups is quantitative analysis of nuclear Overhäuser enhancement (NOE) values.[47,48]

For additional conformational information, a series of variable temperature, one-dimensional NMR experiments is useful in determining the availability of amide protons for solvent exchange.[49,50] If an amide bond proton is solvent-shielded or intramolecularly hydrogen-bonded, its chemical shift will not change or change only very slightly ($\Delta\delta/\Delta T = 0$ to -3×10^{-3} ppm/degree) as a function of temperature. If, on the other hand, an amide proton is not intramolecularly hydrogen-bonded, then its chemical shift will change in direct relation with temperature. This rate of change is on the order of -4 to -8×10^{-3} ppm/degree.

2. Application of NMR to the Analysis of Analogs of α-MSH

NMR investigations of melanotropic peptides have examined both the backbone- and side-

SER-TYR-SER-MET-GLU-HIS-PHE-ARG-TRP-GLY-LYS-PRO-VAL-GLY-LYS-LYS-
5 10 15

ARG-ARG-PRO-VAL-LYS-VAL-TYR-PRO-ASN-GLY-ALA-GLU-ASP
20 25

FIGURE 4. Amino acid sequence of ACTH(1-29).

chain topologies of linear melanotropic peptides. Specifically, fragments of adrenocorticotropic hormone (ACTH) (Figure 4) (ACTH[1-32]), ACTH[1-24]), ACTH[1-10], ACTH[4-10], ACTH[6-10], and ACTH[7-10])[26-30] and α-MSH analogs (4-11 fragments) were examined. Among these systems, the corresponding D-Phe[7] analogs of ACTH(1-10), ACTH(4-10), and ACTH(7-10)[27] were investigated, as well as analogs with individual D-amino acid substitutions in ACTH(6-10)[28,29] where each L-amino acid residue was replaced by its D-amino acid isostere. Also examined were the spectral properties of the individual D-amino acid substitutions in the 6-9 core sequence of Ac-[Nle^4]α-MSH_{4-11}-NH_2.[51]

a. Early Carbon 13 Studies

The conformational features of the largest ACTH analogs, ACTH(1-32) and ACTH(1-24), were studied in an aqueous environment by ^{13}C-NMR, in which the peptides were isotopically enriched in the 7 and 23 positions with the ^{13}C-nucleus (see Table 3).[30] The major objective of this work was to examine possible *cis/trans* isomerism of Pro^{24} and its effect on the surrounding amino acid residues. The percentages of *cis* and *trans* Pro isomers were determined by measurement of the relative intensities of the quaternary aromatic carbons of the adjacent Tyr^{23} in which the carbon chemical shifts were shown to be sensitive to the stereochemistry of the α-carbon in Pro.

Also observed were short- and long-range chemical shift effects in the sequence -Tyr^{23}-Pro^{24}-Asn^{25}-Gly^{26}-Ala^{27}- as induced by the isomerism of Pro. This observation was reported to be indicative of a nonrandom conformation of the peptide in this region under aqueous conditions. Additional ^{13}C-NMR chemical shift effects observed in this region for Gly^{26}, Pro^{24}, and Val^{22} were used to suggest the existence of a conformational equilibrium involving a β-turn in the 23-26 amino acid sequence.[30] Finally, the observation of chemical shift variation for the α- and β-carbons of Phe^7 in going from smaller analogs of ACTH (ACTH[5-14]) to progressively larger analogs (ACTH[1-24] and ACTH[1-32]) was used to suggest the existence of an α-helical structure in the N-terminal portion of the molecule (positions 3-9).[30]

b. Proton NMR Analysis of ACTH Analogs Related to α-MSH

A far more extensive effort has been applied to the analysis of shorter fragments of ACTH, as well as modified ACTH analogs which have D-amino acid substitutions.[28,29] Some of the earlier NMR work was conducted on the ACTH pentapeptide, His-Phe-Arg-Trp-Gly, ACTH(6-10), where each L-amino acid residue was replaced by its respective D-enantiomer (see Figure 1). These studies were prompted by the dramatic effect of a D-amino acid substitution on the melanotropic activity of ACTH fragments.

Individual proton NMR resonances were assigned by selective decouplings and by comparison of the spectra with those of similar peptides. To obtain information about the backbone conformation of the all L- and individually substituted D-amino-acid-containing pentapeptides, the vicinal ($^3J_{NH\text{-}CH}$) coupling constants were examined in denterated dimethyl sulfoxide solution and found to be in the range of 7 to 9 Hz (Table 4). These values are larger than those anticipated for either an α-helix (~5 Hz) or a random coil (~6 to 7 Hz), but rather are closer to the anticipated $^3J_{NH\text{-}CH}$ coupling constants for an extended (180°) or β-type (130°) conformation.[29]

Table 3
^{13}C CHEMICAL SHIFT ASSIGNMENTS FOR ACTH(1-14)[a] IN D_2O SOLUTION (REFERENCE IS DIOXANE AT 67.4 ppm)

Carbon		Chemical shift	Carbon		Chemical shift
Ser^1	α	56.3	Arg^8	α	52.1
	β	61.0		β	27.4
				γ	24.9
Tyr^2	α	55.1		δ	41.3
	β	37.1		ε	157.1
Ser^3	α	56.3	Trp^9	α	55.8
	β	61.8		β	29.0
Met^4	α	54.2	Gly^{10}	α	43.1
	β	31.0			
	γ	30.1	Lys^{11}	α	53.2
	ε	N.D.		β	31.1
					23.6
Glu^5	α	54.2		δ	27.2
	β	26.8		ε	40.1
	γ	N.D.	Pro^{12}	α	61.2
His^6	α	53.8		β	30.2
	β	26.9		γ	25.5, 21.3[b]
				δ	N.D.
Phe^7	α	55.2	Val^{13}	α	60.2
	β	38.1		β	30.8
				γ	19.2, 18.5[c]
			Gly^{14}	α	42.0

[a] ^{13}C chemical shift assignments are available for ACTH(8-14), ACTH(5-14), ACTH(1-14), ACTH(15-24), ACTH(1-24), ACTH(22-26), ACTH(20-32), and ACTH(1-32) in Reference 30.

[b] Corresponds to the *trans* and *cis* isomers of Pro, respectively.

[c] ^{13}C chemical shifts for both of the α-protons of Val 13.

From Toma, F., Fermandjian, S., Low, M., and Kisfaludy, L., *Biopolymers*, 20, 901, 1981. With permission.

Table 4
$^3J_{NH\text{-}CH}$ COUPLING CONSTANTS[a] FOR ACTH(6-10) PENTAPEPTIDES IN DMSO-d_6

Analog	Phe	Arg	Trp
His-Phe-Arg-Trp-Gly	—	7.9	8.8
D-His-Phe-Arg-Trp-Gly	—	7.7	8.7
His-D-Phe-Arg-Trp-Gly	8.3	7.3	9.4
His-Phe-D-Arg-Trp-Gly	8.2	7.7	9.1
His-Phe-Arg-D-Trp-Gly	—	7.9	8.2

[a] Errors related to the digital resolution are ±0.2 Hz in the value reported for the coupling constant.

From Higuchi, N., Kyogoku, Y., and Yajima, H., *Biopolymers*, 20, 2203, 1981. With permission.

The temperature dependencies for the all L-amino-acid-containing pentapeptides and the D-Phe[7]- or D-Arg[8]-containing analogs, were not consistent with a hydrogen-bond-stabilized turn. The amide NH temperature dependencies were all in the normal range of −4.0 to -7.5×10^{-3} ppm/degree (not shown).[29] It was observed that for the D-Phe[7]-containing analog the Trp indole NH proton and Arg epsilon NH proton existed as an unsymmetrical doublet, and other amide protons and α-proton signals were also more complex than expected. In addition, the D-Arg[8]-containing analog also displayed a doublet for the $N^{\epsilon}H$ proton of the Arg side-chain. It was also noted that the β-hydrogen signals were observed as doublets for the D-Phe[7] and D-Arg[8] analogs which collapsed to singlets, when the temperature was raised from 25 to 70°C. These results were interpreted as being indicative of possible steric hindrance of the mobility of individual amino acid side-chains on the NMR time-scale.

Additional evidence for slowly exchangeable conformations on the NMR time-scale was found in the C^{13}-NMR experiments in dimethyl sulfoxide solution where doubling of the peaks was reported for the D-Phe[7] and D-Arg[8] residues in the proton-decoupled spectra of His-D-Phe-Arg-Trp-Gly and His-Phe-D-Arg-Trp-Gly, respectively. The former analog had extremely short T_1 relaxation times for the β-, δ-, and γ- carbons of the Arg side-chain.[29]

ACTH(1-10) and related shortened fragments have also been examined by NMR in D_2O (Table 5) and D_2O/trifluoroethanol (30/70) solutions at pH 2.0 and 7.0[27] in an effort to determine the effect of the D-Phe[7] substitution on the overall conformation of the peptide. Included in this series was ACTH(7-10), ACTH(4-10), and ACTH(1-10) as well as the corresponding D-Phe[7]-containing analogs of each. In comparing the proton NMR spectra in this series, no significant change was observed in the chemical shifts of the core amino-acid residues with the extension of the 7-10 sequence to the 4-10 and 1-10 sequences in either the L-Phe[7] or D-Phe[7] series at either pH. This suggests that extension of the C-terminus from the 7-10 core sequence does not influence the three-dimensional aqueous solution conformation of this peptide.[27] In aqueous solution, the $^3J_{NH\text{-}\alpha CH}$ constants were relatively large (5.5 to 8.5 Hz), consistent with a random coil or β-type structure. The vicinal coupling constants measured in D_2O/trifluoroethanol solutions were generally lower than those found in D_2O (5.0 to 7.6 2Hz), but none was indicative of an α-helix.[27]

The NH chemical shift temperature coefficients were large (-5.0 to -7.0×10^{-3} ppm/degree) in aqueous solution, while the temperature dependencies of the NH proton resonances in D_2O/trifluoroethanol solutions revealed a larger range (-3.7 to -7.8×10^{-3} ppm/degree), but still were not indicative of intramolecular hydrogen bonding or significant secondary structure.

A useful method for addressing side-conformations comes from an estimation of the individual rotamer populations (defined by χ_1) about the α-β-carbon-carbon bond (Figure 5). The differential coupling of the α-proton to the β-protons as a function of spatial orientation was used to calculate the relative ratios of P_I, P_{II}, and P_{III} (trans, gauche [−], and gauche [+])[27] by utilizing the Pachler equation or a modified derivative thereof.[45,46,52]

For this series of ACTH peptides in D_2O, the required coupling constant data could be extracted only for the Phe and Trp residues. A survey of the data showed that in the all L-amino-acid-containing- and the D-Phe[7]-containing series the only significant difference was seen in the side-chain populations of the Phe[7] and Trp[9] residues in the ACTH(7-10) fragment, where Phe was at the N-terminus (Table 6). The only conclusion that can be drawn from this study is that ACTH(1-10) does not form a dominant well-defined conformation in either aqueous or D_2O/trifluoroethanol solutions.[27]

c. NMR Analysis of Linear 4-11 Analogs of α-MSH

An investigation of the aqueous solution conformation of a series of diastereometric analogs of Ac-[Nle[4]]α-$MSH_{4\text{-}11}$-NH_2 and Ac-[Nle[4],D-Phe[7]] α-$MSH_{4\text{-}11}$-NH_2 containing D-isomeric replacements at positions 6, 8 and 9 also concluded that the predominant backbone confor-

Table 5
PROTON CHEMICAL SHIFTS FOR ACTH(1-10) AND [D-Phe[7]]ACTH(1-10) AT pH 2 AND 7 in D_2O;[a,b]

	Residue	ACTH(1-10)		[D-Phe[7]]ACTH(1-10)	
		pH 7	pH 2	pH 7	pH 2
Ser[1]	$C_\alpha H$	—	4.08	—	4.07
	$C_\beta H$	—	3.93	—	3.93
Tyr[2]	$C_\alpha H$	—	4.62	—	4.62
	$C_\beta H$	—	2.96	—	2.97
	$C_{3,5}H$	6.77	6.78	6.76	6.77
	$C_{2,6}H$	7.07	7.07	7.04	7.07
Ser[3]	$C_\alpha H$	—	4.35	—	4.35
	$C_\beta H$	—	3.76	—	3.75
Met[4]	$C_\alpha H$	—	4.37	—	4.37
	$C_\beta H$	—	2.02	—	1.94
	$C_\gamma H$	—	2.50	2.46	2.46
	$C_\delta H$	—	2.04	2.00	2.02
Glu[5]	$C_\alpha H$	—	4.17	—	4.18
	$C_\beta H$	—	1.84	1.77	1.81
	$C_\gamma H$	—	2.30	—	2.25
His[6]	$C_\alpha H$	—	4.54/4.47	—	4.62/4.47
	$C_\beta H$	—	3.06, 3.01/2.81	—	
	C_2H	7.80	8.49	7.84	8.41
	C_4H	6.80	7.11	6.82	7.04
Phe[7]	$C_\alpha H$	—	4.47/4.54	4.48	4.47/4.62
	$C_\beta H$	—	2.81/3.06/3.01	—	—
	Aromatics	7.16	7.20	7.25	7.28
		7.08	7.09	7.14	7.18
Arg[8]	$C_\alpha H$	—	4.17	—	4.07
	$C_\beta H$	—	1.57	1.41, 1.33	1.41, 1.27
	$C_\gamma H$	—	1.34	1.07	0.96
	$C_\delta H$	—	3.05	—	2.88
Trp[9]	$C_\alpha H$	—	4.67	—	4.69
	$C_\beta H$	—	3.30, 3.17	3.31, 3.15	3.29, 3.15
	C_2H	7.22	7.21	7.18	—
	C_4H	7.62	7.61	7.63	7.58
	C_5H	7.11	7.14	7.08	7.08
	C_6H	7.17	7.19	7.13	7.16
	C_7H	7.39	7.37	7.38	7.41
Gly[10]	$C_\alpha H$	—	3.87	—	3.88

[a] Where no value is entered, no value was reported in Reference 27. Where two values are separated by a slash (/), this indicates an uncertainty in the assignment. Where two values are separated by a comma (,), this refers to two independently separated protons (i.e., $C_{\beta(1)}$ and $C_{\beta(2)}$).

[b] Values are reported to be accurate to ±0.015 ppm.

mation was a nonhydrogen-bonded β-like structure.[51] In this study, the anisotropic influences of the aromatic side-chain residues of His[6], Phe,[7] and Trp[9] were utilized to analyze the side-chain topology of these analogs. The wide range of melanotropic potencies of these analogs (32,000 in the frog skin) did not correlate with lipophilicity, suggesting that both a unique backbone conformation and side-chain topology were essential for melanotropic potency. Both Ac-[Nle[4],D-Phe[7]]α-MSH_{4-11}-NH_2, I, (relative potency to the all L-octapeptide = 373) and Ac-[Nle[4],D-Phe[7],D-Trp[9]]α-MSH_{4-11}-NH_2, II, (relative potency = 1668) have a left-handed turn going from C^6_α to C^9_α and a close (i.e., gauche) side-chain topology for His[6], Phe[7], and Arg[8].

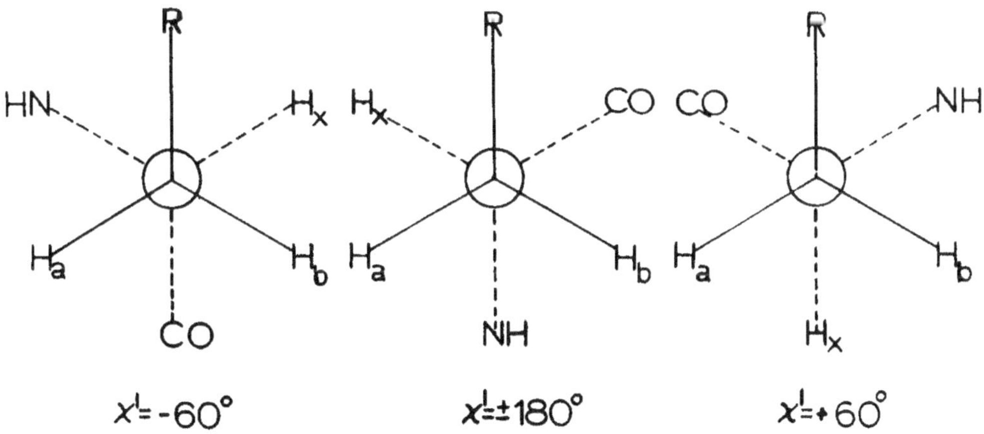

FIGURE 5. Illustration of the three possible rotamer populations: (gauche [+], gauche [−], and *trans*).

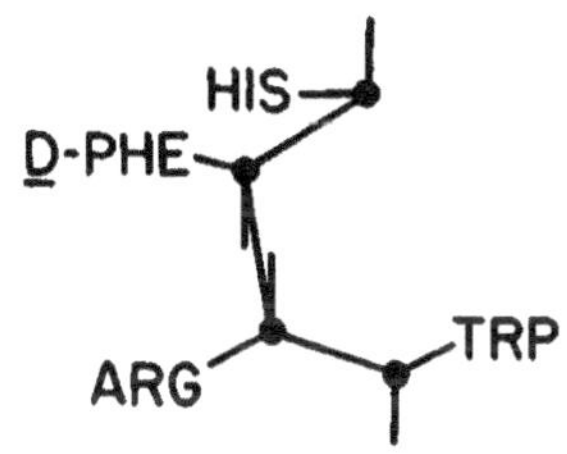

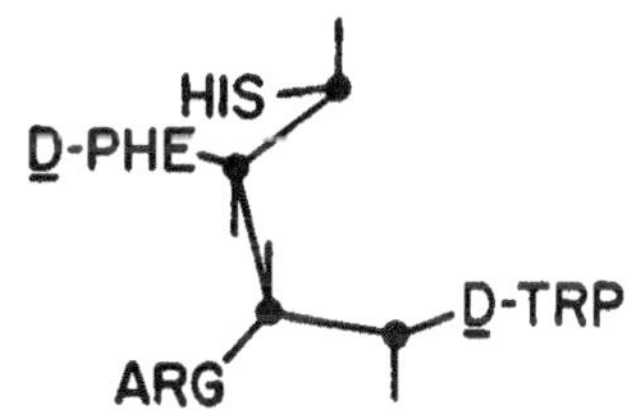

Ac-[Nle⁴,D-Phe⁷]-α-MSH₄₋₁₁-NH₂

I II

Ac-[Nle⁴,D-His⁶,D-Phe⁷]α-MSH$_{4\text{-}11}$-NH$_2$, III, (relative potency = 0.32) has the left-handed turn from C_α^6 to C_α^9, but the His⁶ and Phe⁷ side-chains are *trans* to one another.

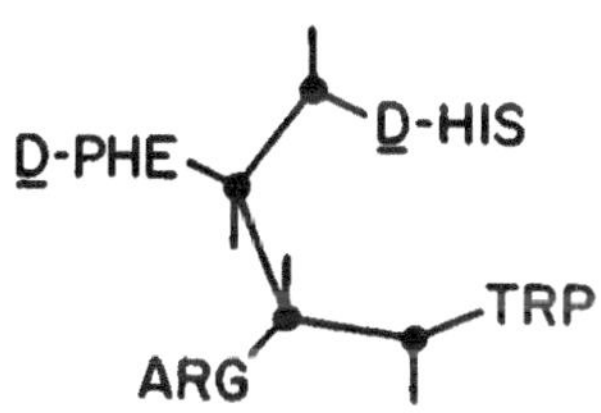

III

Finally, Ac-[Nle⁴,D-Arg⁸]α-MSH$_{4\text{-}11}$-NH$_2$, IV, (relative potency = 0.01) has a right-handed turn from C_α^6 to C_α^9 and a *trans* relationship for the side-chains of His⁶ and Phe⁷.

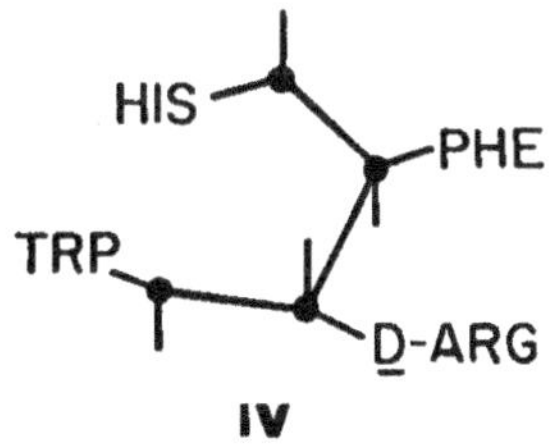

IV

Table 6
THE $J_{\alpha\beta}$ COUPLING CONSTANTS AND CORRESPONDING ROTAMER[b] POPULATIONS FOR Phe^7 AND Trp^9 IN ACTH(7-10) AND [D-Phe^7]ACTH(7-10)

Residue		ACTH(7-10) pH 2	ACTH(7-10) pH 7	[D-Phe^7]ACTH(7-10) pH 2	[D-Phe^7]ACTH(7-10) pH 7
Phe^7	$J_{\alpha\beta1}$	7.7	7.7	9.85	10.1
	$J_{\alpha\beta2}$	6.7	7.5	5.55	5.5
	$J_{\beta\beta}$	13.6	14.1	13.2	13.3
	P_I	0.40	0.45	0.27	0.27
	P_{II}	0.46	0.46	0.66	0.68
	P_{III}	0.14	0.09	0.07	0.05
Trp^9	$J_{\alpha\beta1}$	8.5	8.0	9.8	8.5
	$J_{\alpha\beta2}$5.7	6.3	5.9	7.0	
	$J_{\beta\beta}$	14.7	15.1	14.9	14.6
	P_I	0.28	0.34	0.30	0.40
	P_{II}	0.54	0.49	0.65	0.54
	P_{III}	0.18	0.17	0.05	0.06

[a] Errors in coupling constants were reported to be ± 0.25 Hz.
[b] Errors in rotamer populations are reported to be ± 0.1 and were calculated using the method of Pachler.[52]
[c] P_I = trans, P_{II} = gauche (−), and P_{III} = gauche (+).

III. SUMMARY AND FUTURE PERSPECTIVES

Efforts to understand the relationships of conformation and topology to biological activity of α-melanotropin utilizing physical and molecular modeling methods are still in an elementary state. The studies reported thus far do indicate that even the aqueous solution α-MSH and its analogs can assume preferred conformations, but there is still much to be done to determine with greater certainty and clarity the precise nature of these conformational and topological preferences.

Linear and cyclic analogs of α-MSH, some only one half the size of the native hormone, have been developed which have exquisitive receptor selectivity, superpotency, and exceptionally prolonged in vitro and in vivo biological activities. These findings offer considerable incentive to develop conformational structure models which might explain these biological results, and then to test these models with appropriate conformational constraints consistent with the biological and biophysical data. It can be anticipated that a combination of design and synthesis based on conformational and topological considerations, modeling utilizing molecular mechanics, and molecular dynamics approaches, and biophysical studies (especially NMR) on constrained analogs, will provide significant new insights into α-MSH conformation-activity relationships.

REFERENCES

1. **Hadley, M. E. and Bagnara, J. T.,** Regulation of release and mechanism of action of MSH, *Am. Zool.*, 15 (Suppl. 1), 81, 1975.
2. **Hadley, M. E., Heward, C. B., Hruby, V. J., Sawyer, T. K., and Yang, Y. C. S.,** Biological actions of melanocyte-stimulating hormone, in *Peptides of the Pars Intermedia,* Evered, D. and Lawrenson, G., Eds., Pitman Medical, London, 1981, 244.
3. **LaHoste, G. J., Olson, G. A., Kastin, A. J., and Olson, R. D.,** Behavioral effects of melanocyte-stimulating hormone, *Neurosci. Biobehav. Rev.*, 4, 9, 1980.
4. **O'Donohue, T. L. and Dorsa, D. M.,** The opiomelanotropinergic neuronal and endocrine systems, *Peptides,* 3, 353, 1982.
5. **O'Donohue, T. L., Handelmann, G. E., Chaconas, T., Miller, R. L., and Jacobowitz, D. M.,** Evidence that N-acetylation regulates the behavioral activity of α-MSH in the rat and human central nervous system, *Peptides,* 2, 333, 1981.
6. **Sawyer, T. K., Cody, W. L., Knittel, J. J., Hruby, V. J., Hadley, M. E., Hirsch, D. M., and O'Donohue, T. L.,** Design of conformationally restricted cyclic α-melanotropins: comparison of melanocyte stimulating and behavioral activities, in *Peptides: Structure and Function: Proc. 8th Am. Peptide Symp.*, Hruby, V. J. and Rich, D. H., Eds., Pierce Chemical Company, Rockford, Ill., 1983, 323.
7. **Hruby, V. J., Wilkes, B. C., Cody, W. L., Sawyer, T. K., and Hadley, M. E.,** Melanotropins: structural, conformational and biological considerations in the development of superpotent and superprolonged analogs, in *Peptide and Protein Reviews,* Vol. 3, Hearn, M. T. W., Ed., Marcel Dekker, New York, 1984, 1.
8. **Medzihradszky, K.,** The bio-organic chemistry of α-melanotropin, *Med. Res. Rev.*, 2, 247, 1982.
9. **Sawyer, T. K., Sanfilippo, P. J., Hruby, V. J., Engel, M. H., Heward, C. B., Burnett, J. B., and Hadley, M. E.,** [4-Norleucine,7-D-phenylalanine]-α-melanocyte-stimulating hormone: a highly potent α-melanotropin with ultralong biological activity, *Proc. Natl. Acad. Sci. U.S.A.*, 77, 5754, 1980.
10. **Wilkes, B. C., Sawyer, T. K., Hruby, V. J., and Hadley, M. E.,** Comparative biological activities of potent active-site analogues of α-melanotropin, *Intl. J. Peptide Protein Res.*, 23, 621, 1984.
11. **Cody, W. L., Mahoney, M., Knittel, J. J., Hruby, V. J., de L. Castrucci, A. M., and Hadley, M. E.,** Cyclic melanotropins. IX. D-phenylalanine[7] analogues of the active site sequence, *J. Med. Chem.*, 28, 583, 1984.
12. **Hruby V. J., Cody, W. L., Wilkes, B. C., Rockway, T. W., and Hadley, M. E.,** Design and synthesis of peptide hormones with prolonged activity at specific receptors, in *Peptides 1984, Proc. 18th Eur. Peptide Symp.*, Ragnarsson, U., Ed., Almquist & Wiksell, Stockholm, 1984, 505.
13. **Sawyer, T. K., Hruby, V. J., Hadley, M. E., and Engel, M. H.,** α-Melanocyte-stimulating hormone: chemical nature and mechanism of action, *Am. Zool.*, 23, 529, 1983.
14. **Sawyer, T. K., Darman, P. S., Hruby, V. J., and Hadley, M. E.,** Conformation-activity studies of α-melanocyte-stimulating hormone: synthesis of a cyclic α-melanotropin exhibiting high potency and full biological activity, in *Peptides: Synthesis-Structure-Function, Proc. 7th Am. Peptide Symp.*, Rich, D. H. and Gross, E., Eds., Pierce Chemical Company, Rockford, Ill., 1981, 387.
15. **Sawyer, T. K., Darman, P. S., Hruby, V. J., and Hadley, M. E.,** [4-Half-cystine, 10-half-cystine]-α-melanocyte stimulating hormone: a cyclic α-melanotropin exhibiting superagonistic biological activity, *Proc. Natl. Acad. Sci. U.S.A.*, 79, 1751, 1982.
16. **Knittel, J. J., Sawyer, T. K., Hruby, V. J., and Hadley, M. E.,** Structure-activity studies of highly potent [Cys[4], Cys[10]]melanotropin analogues, *J. Med. Chem.*, 28, 125, 1983.
17. **Venkatachalam, C. M.,** Stereochemical criteria for polypeptides and proteins. V. Conformation of a system of three linked peptide units, *Biopolymers,* 6, 1425, 1968.
18. **Sawyer, T. K.,** α-Melanotropin: Chemical Nature and Mechanism of Biological Activity, Ph.D. thesis, University of Arizona, Tucson, 1981.
19. **Hruby, V. J.,** Conformational restrictions of biologically active peptides via amino acid side chain groups, *Life Sci.*, 31, 189, 1982.
20. **Lebl, M., Cody, W. L., and Hruby, V. J.,** Cyclic melanotropins. Part VI. Reverse phase HPLC studies, *J. Liq. Chromatogr.*, 7, 1195, 1984.
21. **Cody, W. L., Wilkes, B. C., and Hruby, V. J.,** Reversed-phase high-performance liquid chromatography studies of α-MSH fragments, *J. Chromatogr.*, 314, 313, 1984.
22. **Ikeda, K., Urano, W., Hamaguchi, K., Kubo, K., and Yajima, H.,** Circular dichroism of active fragments of α-melanocyte-stimulating hormone, *Chem. Pharm. Bull.*, 19, 1815, 1971.
23. **Nikiforovich, G., Shenderovich, M. D., and Chipens, G.,** The space structures of α-melanotropin, *FEBS Lett.*, 126, 180, 1981.
24. **Nikiforovich, G., Rozenblit, I., Liepina, I., and Chipens, G.,** Conformational calculations and biologically active conformations of cyclopeptides, in *Peptides 1982, Proc. 17th Eur. Peptide Symp.*, Blaha, K. and Malon, P., Eds., Walter de Gruyter, Berlin, 1983, 735.

25. **Nikiforovich, G. V., Rozenblit, M. D., Shenderovich, M. D., and Chipens, G. I.,** Possible bioactive conformations of α-melanotropin, *FEBS Lett.*, 170, 315, 1984.
26. **Patel, D. J.,** Proton nuclear magnetic resonance study of corticotropins, *Macromolecules*, 4, 251, 1971.
27. **Rawson, B. J., Feeney, J., Kimber, B. J., and Greven, H. M.,** ^{1}H nuclear magnetic resonance studies of the conformations of adrenocorticotropic hormone ACTH(1-10) and related peptides in aqueous and trifluoroethanol solutions, *J. C. S. Perkin Trans.*, II, 1471, 1982.
28. **Higuchi, N., Kyogoku, Y., and Yajima, H.,** Nuclear magnetic resonance studies of pentapeptides with weak ACTH activity and their solution conformations, in *Peptide Chemistry 1979*, Yanihara, H., Ed., Protein Research Foundation, Osaka, Japan, 1980, 159.
29. **Higuchi, N., Kyogoku, Y., and Yajima, H.,** Conformational studies of the pentapeptides with weak adrenocorticotropin (ACTH) activities by means of nuclear magnetic resonance spectroscopy, *Biopolymers*, 20, 2203, 1981.
30. **Toma, F., Fermandjian, S., Low, M., and Kisfaludy, L.,** ^{13}C-NMR studies of ACTH: assignment of resonances and conformational features, *Biopolymers*, 20, 901, 1981.
31. **Lebl, M.,** Amino acids and peptides CLXXVIII. Correlation between the hydrophobicity of substituents in the phenylalanine moiety of oxytocin carba-analogues and reversed-phase-chromatography k′ values, *J. Chromatogr.*, 292, 392, 1982.
32. **Woody, R. W.,** Circular dichroism of peptides, in *Peptides: Analysis, Synthesis, Biology*, Vol. 7, Hruby, V. J., Ed., Academic Press, New York, 1985, 16.
33. **Scheraga, H. A.,** Theoretical studies of molecular recognition and catalysis by enzymes, in *Design and Synthesis of Organic Molecules Based on Molecular Recognition, Proc. 18th Solvay Conf. on Chem.*, Van Binst, G., Ed. Springer-Verlag, New York, 1983, 117.
34. **Ramachandran, G. N., Ramakrishnan, C., and Sasisekharan, V. J.,** Stereochemistry of polypeptide chain configurations, *Mol. Biol.*, 7, 95, 1963.
35. **Ramachandran, G. N. and Sasisekharan, V.,** Conformation of polypeptides and proteins, *Adv. Protein Chem.*, 23, 283, 1968.
36. **Scheraga, H. A.,** Calculations of conformations of polypeptides, in *Advances in Physical Organic Chemistry*, Vol. 6, Gold, V., Ed., Academic Press, New York, 1968, 103.
37. **Hagler, A. T.,** Theoretical simulation of conformation, energetics and dynamics of peptides, in *The Peptides: Analysis, Synthesis, Biology*, Vol. 7, Hruby, V. J., Ed., Academic Press, New York, 1985, 214.
38. **Dauber, P., Goodman, M., Hagler, A. T., Osguthorpe, D., Sharon, R., and Stern, P.,** Computer simulation of biomolecular systems in supercomputers in chemistry, in *ACS Symp. Ser. 173*, Lykos, P. and Shavitt, I., Eds., American Chemical Society, Washington, D. C., 1981, 161.
39. **Abillon, E., Chuong, P. V., and Fromageot, P.,** Conformational calculations on gastrin C-terminal tetrapeptide, *Intl. J. Peptide Protein Res.*, 17, 480, 1981.
40. **Struthers, R. S., Rivier, J., and Hagler, A. T.,** Molecular dynamics and minimum energy conformations of GnRH and analogs: a methodology for computer-aided drug design, in *Macromolecular Structure and Specificity: Computer-Assisted Modeling and Applications*, Venkatarughavan, B. and Feldmann, R. V., Eds., *Ann. N.Y Acad. Sci.*, 439, 81, 1984.
41. **Akhmedov, N. A., Adhverdieva, G. A., Godjaev, N. M., and Popov, E. M.,** Theoretical conformational analysis of brain peptides, β-melanocyte-stimulating hormone, *Intl. J. Peptide Protein Res.*, 27, 95, 1986.
42. **Kessler, H., Bermel, W., W., Müller, A., and Pook, K.-H.,** Modern nuclear magnetic resonance spectroscopy of peptides, in *The Peptides: Analysis, Synthesis, Biology*, Hruby, V. J., Ed., Academic Press, New York, 1985, 438.
43. **Morris, G. A.,** Modern NMR techniques for structure elucidation, *Magn. Reson. Chem.*, 24, 371, 1986.
44. **Kessler, H.,** Detection of hindered rotation and inversion by NMR spectroscopy, *Angew. Chem. Int. Ed.*, 9, 219, 1970.
45. **Bystrov, V. F., Portnova, S. L., Tsetlin, V. I., Ivanov, V. T., and Ovchinnikov, Y. A.,** Conformational studies of peptide systems. The rotational states of the NH-CH fragment of alanine dipeptides by nuclear magnetic resonance, *Tetrahedron*, 25, 493, 1969.
46. **de Leeuw, F. A. A. M. and Altona, C.,** Component vicinal coupling constants for calculating side-chain conformations in amino acids, *Intl. J. Peptide Protein Res.*, 20, 120, 1982.
47. **Jones, C. R., Sikakana, C. T., Hehir, S., and Chang, M.,** The quantitation of nuclear Overhäuser effect methods for total conformational analysis of peptides in solution: application to Gramicidin S, *Biophys. J.*, 24, 815, 1978.
48. **Roques, B. P., Rao, R., and Marion, D.,** Use of nuclear Overhäuser effect in the study of peptides and proteins, *Biochemie*, 62, 753, 1980.
49. **Gibbons, W. A., Nemethy, G., Stern, A., and Craig, L. C.,** An approach to conformational analysis of peptides and proteins in solution based on a combination of nuclear magnetic resonance spectroscopy and conformational calculations, *Proc. Natl. Acad. Sci. U.S.A.*, 67, 239, 1970.
50. **Urry, D. W., Masotti, L., and Krivacic, J. R.,** Improved ellipicity data for several biological membranes, *Biochem. Biophys. Res. Commun.*, 41, 521, 1970.

51. **Sugg, E. S., Cody, W. L., Abdel-Malek, Z. A., Hadley, M. E., and Hruby, V. J.,** *D*-isomeric replacements within the 6-9 core sequence of Ac-[Nle[4]]-α-MSH_{4-11}-NH_2; a topological model for the solution conformation of α-melanotropin, *Biopolymers,* 25, 2029, 1986.
52. **Pachler, K. G. R.,** Nuclear magnetic resonance studies of some α-amino acids. II. Rotational isomerism, *Spectrochim. Acta,* 20, 581, 1964.

Chapter 8

MELANOTROPIN RECEPTORS: STUDIES WITH LABELED MELANOTROPINS

Alex N. Eberle, Pierre N. E. de Graan, Walter Siegrist, and Jürg Girard

TABLE OF CONTENTS

I. Introduction 112

II. Labeled MSH and MCH Derivatives 112
- A. Tritiated α-MSH and MCH 112
- B. Radioiodinated α- and β-MSH 113
- C. Fluorescent and Macromolecular Labels for MSH 113

III. MSH Derivatives Carrying Affinity Groups 114
- A. α-MSH with Affinity Groups 114
- B. α-MSH with Photoaffinity Groups 115

IV. Characterization and Visualization of MSH Receptors 116
- A. MSH-Receptor-Binding Studies with Melanoma Cells 116
- B. MSH-Receptor Visualization 116
- C. MSH-Receptor Cross-Linking 116

V. Functional Characterization of MSH Receptors Using Photoaffinity Labeling 117
- A. Irreversible Stimulation of MSH Receptors 117
- B. Role of Calcium 119
- C. Effect of Catecholamines 121
- D. Mechanism of Signal Transduction 122

VI. Conclusion 123

Addendum 124

Acknowledgments 124

References 124

I. INTRODUCTION

The mechanism of melanocyte-stimulating hormone (MSH) action has been investigated mainly with extensive structure-activity studies using various assays, such as melanophore and melanoma cell assays or behavioral tests.[1-3] Although details concerning both the organization of hormonal information and the complementary read-out devices in the receptor could be obtained from these studies,[4] information concerning receptor function and characteristics has remained rather limited. Therefore, several years ago, efforts were undertaken to synthesize and apply MSH derivatives containing radioactive, fluorescent, or affinity labels in order to identify, visualize, or tag MSH receptors.

Peptides that retain high biological activity in spite of the modifications produced by the labels are a prerequisite for the study of MSH receptors with labeled MSH derivatives. For example, detailed analysis of structure-activity studies revealed that the central and C-terminal parts of α-MSH, both of which have an important function in receptor stimulation,[5] are more sensitive to structural alterations than the N-terminal part. Bulky affinity or fluorescent labels should, therefore, be introduced into the N-terminus of α-MSH. MSH receptor visualization, identification, and characterization by autoradiography, fluorescence microscopy, photoaffinity labeling, or binding studies not only require biologically highly potent ligands, but also suitable assay systems to which these techniques can be applied. This review will cover the application of the various radioactive MSH derivatives and MSH compounds containing affinity groups, with special emphasis on the photoaffinity labeling of MSH receptors on intact melanophores, a system which yields unique information about receptor function.

II. LABELED MSH AND MCH DERIVATIVES

MSH receptors on melanocytes, melanoma cells, in the brain, and in other tissues, occur in very small numbers and, hence, can only be studied with biologically active MSH derivatives of high specific radioactivity. Isotopes suitable for the MSHs are ^{3}H and ^{35}S, both of which can be introduced isosterically into the molecules. Since, however, ^{35}S would have to be incorporated in the form of [^{35}S]-methionine, chemical preparations are very expensive (an alternative would be to produce the hormones biosynthetically with pituitary cells). If a specific radioactivity higher than that of ^{3}H is required and if the structure of the labeled ligand does not need to be isosteric with the natural hormone, ^{125}I is the label of choice, also because it is more easily incorporated into the peptide.

A. Tritiated α-MSH and MCH

Several tritiated α-MSH derivatives and one tritiated MCH analog have been described in the literature; their structures and specific radioactivities are summarized in Table 1. The first tritiated α-MSH (compound one)[6] was prepared by a conventional synthesis using [^{3}H]Tyr^{2} instead of Tyr^{2}, but the specific activity obtained was very low. In later tritiations of α-MSH, most authors first synthesized a precursor containing either 3′,5′-dibromo- or 3′,5′-diiodotyrosine^{2} for catalytic halogen-tritium exchange (compounds 2 to 7)[7-11] More recently, [$^{3}H_2$]Nle was incorporated into a D-Phe^{7}-containing (4-11) fragment of α-MSH (compound 8),[12] and propargylglycine (Pra) instead of Val^{13} was used for the catalytic insertion of four tritium atoms per molecule (compound 9).[13] This approach was also used in the only synthesis of a tritium-containing MCH analog so far reported: the two methionines at positions 3 and 6 were replaced by Pra (compound 10). The tritiation of the cyclic heptadecapeptide was performed under slight pressure yielding a specific radioactivity of 220 Ci/mmol.[14,15] All these compounds retained a high biological activity and were used for binding and degradation studies.

Table 1
TRITIATED MSH AND MCH DERIVATIVES

Compound	Structure	Specific radioactivity (ci / mmol)	Ref.
1	[(^{3}H)Tyr^2]α-MSH	2.8	6
2	[(^{3}H)Tyr^2]α-MSH	2.6-4	7
3	[D-Ala^1,(^{3}H$_2$)Pap^2,Nva^4]α-MSH	11	8
4	[Bromoacetyl-D-Ala^1,(^{3}H$_2$)Tyr^2,Gly^3,Nva^4]α-MSH	20	9
5	[Diazoacetyl-Gly^1,(^{3}H$_2$)Tyr^2,Nva^4]α-MSH	36	9
6	[(^{3}H$_2$)Tyr^2]α-MSH	34	10
7	[(^{3}H$_2$)Tyr^2]α-MSH	42	11
8	[(^{3}H)Nle^4,D-Phe^7]α-MSH$_{4\text{-}11}$	32	12
9	[(^{3}H$_4$)Nva^{13}]α-MSH	110	13
10	[(^{3}H$_4$)Nva^{3,6}]-MCH	220	14, 15

Note: Nva:norvaline; Pap: *p*-azidophenylalanine.

B. Radioiodinated α- and β-MSH

The preparation of biologically active [^{125}I]-containing MSH tracers requires special precautions since it has been demonstrated by Heward et al.[16-18] that both α- and β-MSH are almost completely inactivated when exposed to the conventional conditions of chloramine T iodination. Therefore, the original study of MSH-receptor binding[19] was probably done with biologically insufficiently characterized β-MSH. The inactivation of the MSHs was thought to be caused largely through oxidation of methionine,[17] but since iodination of [Nle^4]α-MSH and [Nle^7]β-MSH with excess chloramine T also caused loss in biological activity, His and Trp may also have become modified.[17] A more or less equimolar chloramine T iodination procedure, which has previously been shown to yield biologically active normal (i.e., methionine-containing) [^{125}I]α-MSH, also gave biologically active Nle-containing radioiodinated analogs.[18] An alternative to the "equimolar" chloramine T procedure for iodinations of MSHs is the use of iodogen[18] or lactoperoxidase.[20]

The most important factor in the preparation of biologically active radioiodinated MSH is the purification procedure. Although gel filtration through Sephadex® G-25 or ion-exchange chromatography on DEAE-Sephadex® has been widely used, reversed phase high-performance liquid chromatography (RP-HPLC) gives a cleaner separation of the different contaminants from the monoiodinated product. An example of a RP-HPLC separation of [Tyr(I)^2]α-MSH from α-MSH, the diiodo analog, and their oxidized species is presented in Figure 1. This procedure yields monoiodinated α-MSH tracer with biological activity of 50 to 150% of that of α-MSH, depending on the bioassay.[21] A similar separation procedure has also been developed for β-MSH.[22] Recently, a simple purification method consisting of absorption to C18 silica and elution with a step-wise gradient of methanol dilute acid was developed which circumvents HPLC of radioactive material and yields a product of almost the same quality.[21]

The biological testing of radioiodinated MSH requires a bioassay of great sensitivity and small incubation volumes. This is best fulfilled with the new *in situ* melanin assay using B16 mouse melanoma cells in culture (ED$_{50}$: 27 p*M*, i.e., similar to the frog skin assay; incubation volume: 200 μℓ).[23] The receptor binding assay is a further test for biological activity.[24]

C. Fluorescent and Macromolecular Labels for MSH

Instead of using radioactive labels, fluorescent groups,[25] or macromolecules,[1,26,27] or

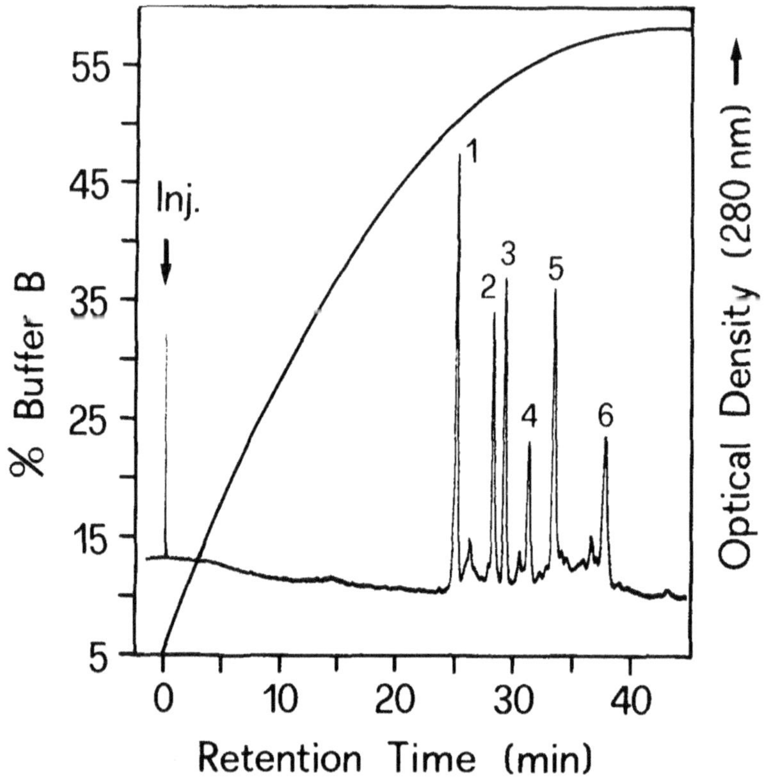

FIGURE 1. Separation by RP-HPLC of [monoiodo-Tyr²]α-MSH (5) from [diiodo-Tyr²]α-MSH (6), α-MSH (3), oxidized α-MSH (1), oxidized [monoiodo-Tyr²]α-MSH (2), and oxidized [diiodo-Tyr²] α-MSH (4).[21] The peptides were eluted from an ODS-silica column with an exponential gradient of the following buffer system; A — 0.1% trifluoroacetic acid/water; B — 0.1% trifluoroacetic acid in acetonitrile/water 7:1.

combinations thereof[28] have been tested for MSH receptor studies. It is not surprising that the introduction of bulky groups into either α- or β-MSH reduces their biological activity and changes their physicochemical properties. The increased lipophilicity of fluorescein-, rhodamine-, or dansyl-containing peptides may produce considerable nonspecific binding, and, hence, these compounds are more prone to give artifacts in receptor visualization than [^{3}H]-labeled peptides used for autoradiography. In this respect, the macromolecule-MSH complexes would seem to be even less suitable than fluorescent peptides. However, since one molecule of carrier protein usually contains several peptide molecules covalently attached to it, the reduced biological activity and changed physiocochemical properties of each individual MSH molecule are compensated for by the presence of more than one peptide molecule on the macromolecular carrier. For example, a tobacco mosaic virus/α-MSH complex containing 300 peptide molecules was 1500 times more potent than the peptide used for covalent attachment or 150 times more potent than natural α-MSH.[1] This increase in potency, which may be explained as a cooperativity phenomenon, could prove useful in receptor localization and isolation, as very stable MSH-receptor complexes can be formed with these conjugates.

III. MSH DERIVATIVES CARRYING AFFINITY GROUPS

A. α-MSH with Affinity Groups

The possible involvement of SH groups in MSH-receptor stimulation reported several

Table 2
α-MSH WITH PHOTOAFFINITY GROUPS

Compound	Structure	MSH activity (*Anolis*)[a] (%)	Labeling efficiency[b] (%)	Ref.
11	[D-Ala[1],Pap[2],Nva[4]]α-MSH	8	65	8
12	[Pap[13]]α-MSH	40	100	33
13	[Apac-Ser[1]]α-MSH	50	100	34
14	[ApSSpr-Ser[1]]α-MSH	33	93	37
15	[Tdbz-Ser[1]]α-MSH	33	100	32
16	[Trp(Naps)[9]]α-MSH	33	100	35
17	[Leu[9],Pap[13]]α-MSH	1	20	36
18	[Apac-Ser[1],Leu[9]]α-MSH	1	68	36

Note: Apac: azidophenylacetyl; ApSSpr: (4-azidophenyl)-1,3′-dithio-propionyl; Naps: 2-nitro, 4-azido-sulfenylphenyl; Pap: *p*-azidophenylalanine; Tdbz: 4-(3-trifluoromethyldiazirino)benzoyl.

[a] Relative to α-MSH (= 100%).
[b] As estimated from the persistence of the long-lasting signal (melanophore dispersion in *Anolis* skin) after a 100-min wash period.

years ago[29] prompted us to investigate α-MSH derivatives containing SH sensitive affinity groups for receptor labeling.[9] Neither derivatives carrying a bromoacetyl or diazoacetyl group in the place of the N-terminal acetyl group nor derivatives with a maleimido group attached to the N^{α}-amino group of Ser^1 or the N^{ϵ}-amino group of Lys,[11] inserted into MSH receptors, as determined with the corresponding radioactive peptides. The compounds were, however, successfully applied to the synthesis of α-MSH-carrier protein complexes used for receptor studies (see above) or the generation of α-MSH antibodies.

A different type of affinity labeled α-MSH was reported by Chaturvedi and co-workers.[30] They prepared a covalent conjugate between α-MSH and biotin which retained a high affinity to avidin. The biotin-avidin system may not only prove useful for receptor isolation, but is also valuable for microscopic receptor identification, as demonstrated with various other systems.

B. α-MSH with Photoaffinity Groups

The generation of a chemically reactive species of MSH ligands generated by photoactivation *in situ*, i.e., after binding to the receptor, has a much better chance of producing receptor labeling than the much more selective chemical affinity labels.[31] A number of photoreactive α-MSH derivatives have been synthesized and studied in labeling experiments with *Xenopus* and *Anolis* melanophores and, in part, with mouse melanoma cells (Table 2). The two types of precursor tested were phenylazide and phenyldiazirine, both of which were equally suitable when the biological system was sufficiently stable for a broad spectrum UV irradiation (compounds 12, 13, and 15).[32] Melanoma cells in culture, however, which are sensitive to the shorter wavelengths of UV light (<310 to 340 nm), are preferably labeled with diazirino compounds because these can more easily be photolyzed by the longer wavelengths of UV light (i.e., 365 nm).[32] Whether the diazirino compounds would result in a higher degree of nonspecific labeling when applied to receptor isolation, has not yet been clarified.

The position of the label within the α-MSH molecule is not very crucial since good labeling efficiency could be obtained with N- and C-terminal photolabels[33,34] as well as labels introduced into the Trp^9 indole ring[35] (compounds 12, 13, 14, and 16). However, the labeling efficiency was much lower with the label in position (compound 11)[8] or when alterations were made in the α-MSH molecule, such as an exchange of Trp^9 by Leu (compounds 17 and 18).[36] Photolabels containing a disulfide bond (e.g., compound 14) were shown to be cleaved after attachment to the receptor when exposed to thiol-containing buffers.[35,37]

The conditions for photolysis of photoreactive α-MSH and the labeling of MSH receptors need to be adjusted to each of the different tissues. Experimental details can be found in the literature.[33,37,38]

IV. CHARACTERIZATION AND VISUALIZATION OF MSH RECEPTORS

A. MSH-Receptor-Binding Studies with Melanoma Cells

The characterization of MSH receptors by quantitative binding studies has only recently been undertaken, although several reports on qualitative and quantitative binding of [^{125}I]-β-MSH to Cloudman S91 melanoma cells appeared several years ago.[19,25,39] This work was criticized because of the iodination procedure used which was destructive to the peptide (see above). In a preliminary study using [^{3}H]-α-MSH as the label, the problem of a large percentage of nonspecific binding could not be resolved.[40] The notion by Pawelek[41] that MSH-receptor binding and, hence, receptor expression are markedly increased or even restricted to the G2 phase of the cell cycle and that functional MSH receptors are internalized into the melanosome, has not been generally accepted.[42]

A binding assay for Cloudman S91 melanoma cells using biologically active [^{125}I]-β-MSH was established by Lambert et al.[43] They found that the pH optimum of the binding was 6.8 and that saturation was reached after a 2- or 3-hr incubation period at 25°C or, respectively, 15 or 0°C. Reversibility of the binding was quantitative after 2 hr at 25°C or 3 to 4 hr at 15°C. Scatchard analysis revealed a K_D of 5 nM for β-MSH and an average number of binding sites per cell of about 6500.[43]

A slightly different assay for B16 melanoma cells using [(^{125}I)Tyr^{2},Nle^{4}]α-MSH was recently established by Siegrist et al.[24] after resolving the problem of peptide degradation and nonspecific binding by the addition of 1,10-phenanthroline. The time-course of the binding differed from that obtained with β-MSH and Cloudman S91 cells, except that at 15°C, saturation was also reached after a 3-hr incubation. Figure 2 depicts a Scatchard analysis of a saturation binding experiment: there was one single class of binding sites with a K_D of 1.7 nM; the number of binding sites per cell was 9750.[24] Structure-activity studies with α-MSH analogs comparing binding, tyrosinase stimulation, and melanin production revealed a close relationship of the activities of the peptides tested in the different assays, except that adrenocorticotropin (ACTH) (1-24) and [Nle^{4}]α-MSH showed a relatively higher binding than melanogenic activity when compared with α-MSH as standard.[24]

Binding studies have also been applied to the characterization of MSH receptors on human melanoma cells.[44] Approximately one half of the tumor cell lines showed significant MSH binding pointing to the existence of MSH receptors on these cells.

B. MSH-Receptor Visualization

The early observations by fluorescence microscopy with [FITC]-β-MSH and autoradiography with [^{125}I]-labeled β-MSH, showed that the hormone binds to the perinuclear area but their internalization into Golgi vesicles[19] have not yet been reinvestigated with biologically well-characterized ligands. A later study with a $(FITC)_6$-ferritin-$(\alpha\text{-MSH})_5$ conjugate showed that binding sites for this complex were restricted to the cell membrane.[28] Quantitative autoradiography in combination with binding studies will soon become a general technique for the visualization of MSH receptors on melanoma tumor slices and later may also resolve the problem of the "life-cycle" of MSH receptors.[61]

C. MSH-Receptor Cross-Linking

Until the advent of reproducible receptor binding assays it was impossible to characterize

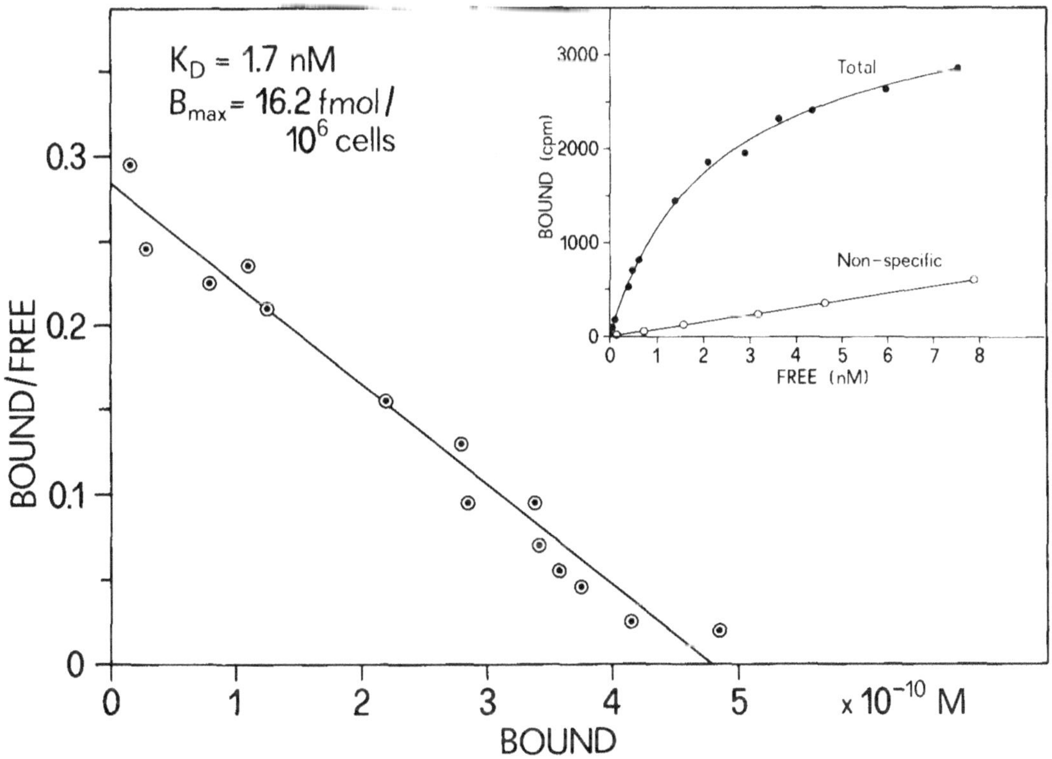

FIGURE 2. Scatchard analysis of a saturation experiment of [^{125}iodo-Tyr^2, Nle^4]α-MSH binding to B16 mouse melanoma cells.[24] Insert: bound tracer per 670,000 cells in relation to increasing ligand concentration. The K_D is 1.7 nM, the B_{max} is 16.2 fmol, and the number of binding sites per cell is 9750.

MSH receptors by cross-linking of [^{125}I]α-MSH or [^{125}I]β-MSH with the receptor and subsequent analysis of the labeled products by sodium dodecylsulfate (SDS) polyacrylamide gel electrophoresis. The recent advent of suitable assays has now prompted several laboratories to tag the receptor; labeled bands with molecular weights ranging from 45 to 120 kdaltons have been reported, but it is too early to draw any conclusion about the real size of the MSH receptor. It is, however, likely that parts of it would be similar to the receptor for ACTH.[45] (See Addendum).

V. FUNCTIONAL CHARACTERIZATION OF MSH RECEPTORS USING PHOTOAFFINITY LABELING

Peptide hormones containing a photoaffinity label can function either as irreversible agonists or as antagonists when used for receptor labeling of intact biological systems.[31] The concept of "irreversible agonism" is particularly useful for studying hormone-receptor interaction and signal transduction in intact cells and provides a new tool for the functional characterization of receptors.[31] Irreversible stimulation by photoaffinity labeling was first described for *Xenopus* melanophores[38] and subsequently for *Anolis*[37] melanophores. Later, the principle was also applied to a variety of other hormonal systems.[31]

A. Irreversible Stimulation of MSH Receptors

When tailfin pieces of *X. laevis* tadpoles or skin pieces of *A. carolinensis* were incubated in the dark with photoreactive α-MSH and then irradiated with UV light (>310 nm), the

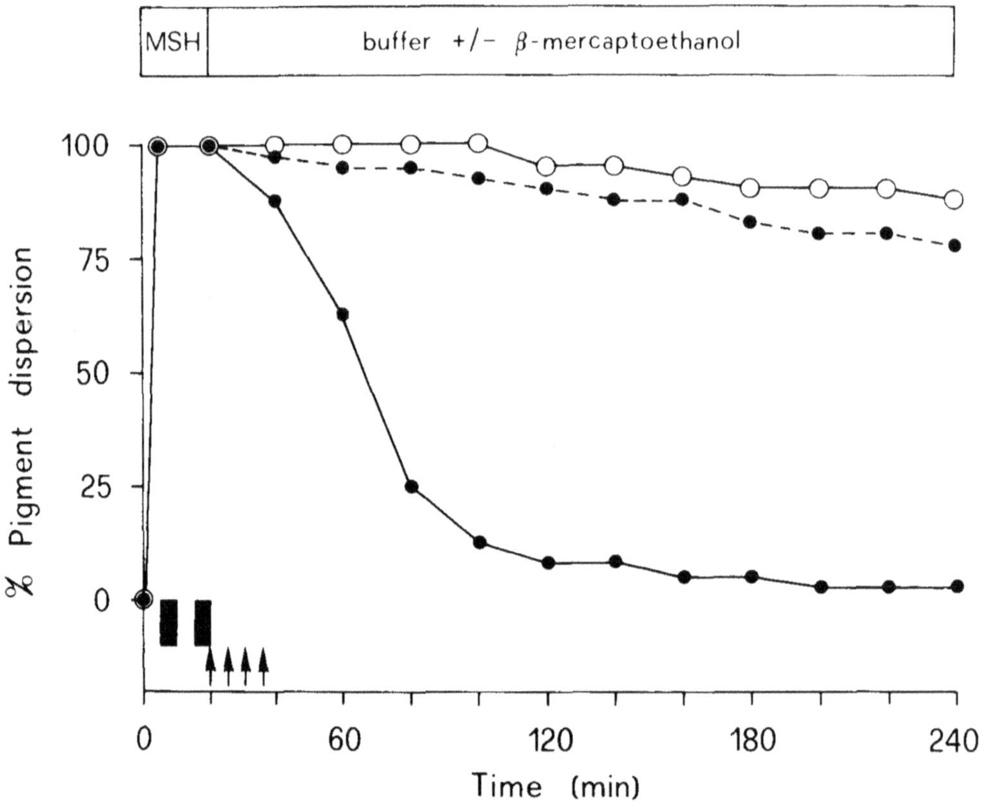

FIGURE 3. Irreversible stimulation of *Anolis* melanophores after formation of a covalent α-MSH-receptor complex by photoaffinity labeling.[37] The skin pieces were incubated in the dark with [Apac-Ser^1]α-MSH (○) or [ApSSpr-Ser^1]α-MSH (●) for 5 min, UV-irradiated for 5 min, and this procedure repeated with fresh peptide solution. The skins were then washed with normal buffer (- - - -) or buffer containing 1% β-mercaptoethanol (——). The long-lasting stimulation is abolished when the hormone containing an S-S-bridged photolabel is cleaved from the receptor by a thiol-containing buffer. (■): UV irradiation phase. (↑): Buffer change. Each point usually represents the mean of 6 to 12 measurements. S.E.M. values are normally 6 to 12%.

pigment dispersion in the melanophores persisted for many hours in spite of continuous washing of the tissues with hormone-free buffer.[37,38] A similar long-lasting receptor activation was also observed when cultured melanoma cells were labeled and subsequently tested for tyrosinase activation.[32,62] The irreversible stimulation of the melanophores has been shown to be specific by a number of tests, such as (1) the blocking of the long-lasting stimulation by the presence of an excess of normal α-MSH during UV irradiation,[38] (2) the inability of α-MSH peptides without photolabel or non-MSH peptides with photolabel to induce irreversible stimulation,[38] (3) the partial reduction of the labeling yield by the presence of the weak antagonist His-Phe-Ala-Trp-Gly-Gly-Pro-Val-NH_2,[37] and (4) the reversibility of the effect in the presence of mercaptoethanol when a disulfide-containing photolabel was used (Figure 3).[37]

The labeling yield did not differ between C- and N-terminal photolabels and ranged from 10 to 25%,[33,38,46] whereas it was only marginally inferior with a photolabel on Trp^9.[37] By varying the wash-out temperature it was possible to estimate receptor turnover: at ≤15°C, covalent MSH-receptor complexes were rather stable when a large proportion of the MSH receptors had been covalently labeled; at 20°C receptor inactivation ranged from 20 to 30%; and at 26°C, from 40 to 60%/hr.[38] If, however, only a small proportion of the MSH receptors was labeled, inactivation could also be observed at 15°C. This inactivation did not depend on the presence or absence of calcium (Figure 4).[34]

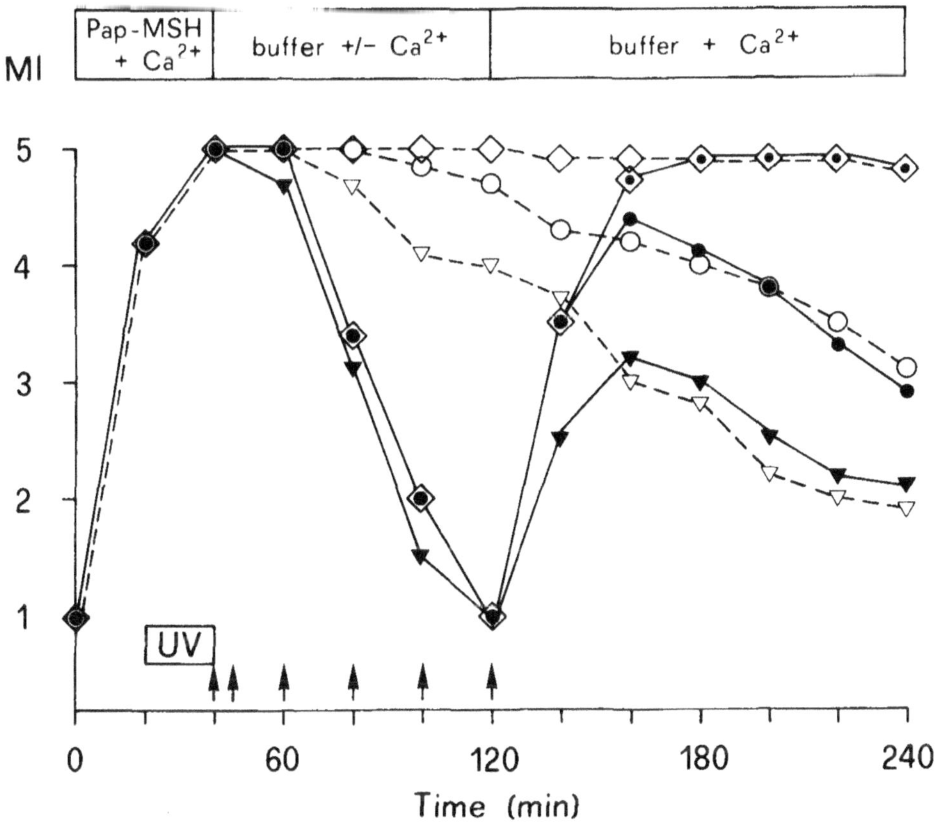

FIGURE 4. MSH-receptor inactivation in irreversibly stimulated *Xenopus* melanophores in the presence and absence of calcium.[34] Photoaffinity labeling was performed in normal medium with 1.2×10^{-8} *M* (◇, ◈), 9×10^{-9} *M* (○, ●), or with 6×10^{-9} *M* (▽, ▼) [Pap^{13}]α-MSH. Then one part of the tailfin pieces (open symbols) was washed in normal buffer. The other part (closed symbols) was washed in calcium-free buffer containing 10^{-4} *M* EGTA to which later (t = 120 min) 2×10^{-3} *M* Ca^{2+} was readded.

A detailed study on how the long-lasting stimulation of *Xenopus* melanophores depended on the hormone concentration used during photolysis is shown in Figure 5. The graph represents log dose-response curves recorded every 20 min during a wash period of 200 min after photolysis. The initial curve is shifted towards approximately a tenfold higher concentration after a 200-min wash period. Although the curves are not parallel, the results permit the conclusion that about one out of ten receptors initially occupied became labeled and that an approximately ten-fold higher hormone concentration needed to be present during photolysis to ensure formation of as many covalent hormone-receptor complexes as there were initially noncovalent ones.[33] Alternatively, the fact that the log dose-response curves became progressively steeper during the 200-min wash period could indicate a threshold concentration for the generation of the long-lasting signal. It is possible that at high hormone concentrations a large number of spare (low-affinity) receptors were labeled which maintained the long-lasting response because of their slow turnover, whereas at low hormone concentration only high-affinity receptors with rapid turnover were labeled.[33,35]

B. Role of Calcium

Calcium is required for hormone-receptor binding and for receptor-effector coupling. In both the *Xenopus* and the *Anolis* melanophore systems, no covalent hormone-receptor complex could be formed in the absence of calcium.[47-49] Similarly, preincubation of *Xenopus*

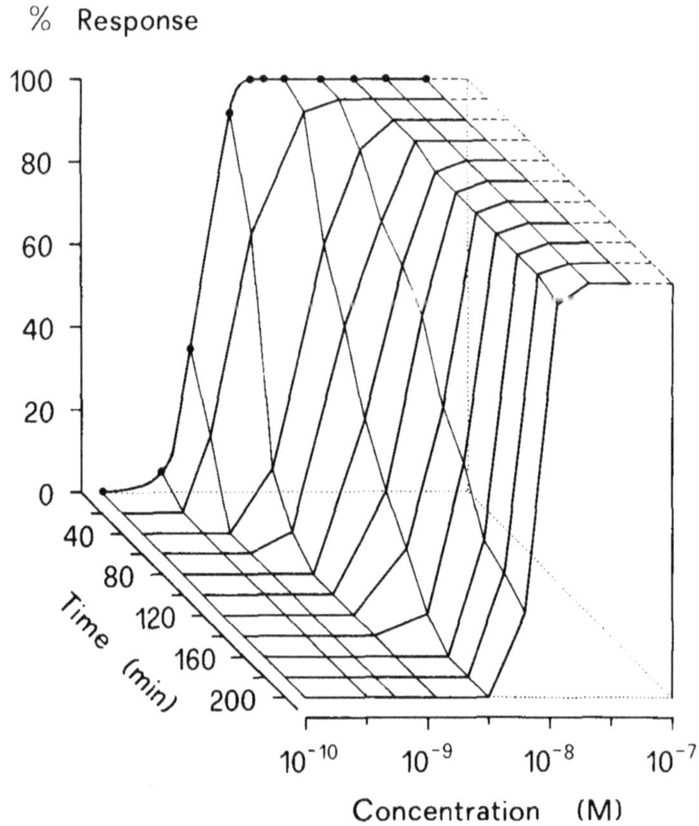

FIGURE 5. Time course of the log dose-response curve after UV irradiation of tailfin melanophores of *Xenopus* tadpoles in the presence of [Pap^{13}]α-MSH.[33,35] The melanophores were labeled with different concentrations of the hormone (20 min preincubation with the peptide, followed by a 20 min UV-irradiation phase). The log dose-response curve was recorded just after photolysis and then every 20 min during a 200-min wash period with buffer.

tailfins with the calcium antagonists, methoxyverapamil (D600; blocks Ca^{2+} influx into cells) or La^{3+} (competitor for Ca^{2+} binding sites on membranes) prevented long-lasting stimulation of the melanophores after UV irradiation with photoreactive α-MSH.[47] When these calcium antagonists were applied after the formation of a covalent hormone-receptor complex, they were no longer effective. However, removal of the calcium from the medium of both *Xenopus* and *Anolis* melanophores led to complete pigment aggregation when EGTA was present (Figure 4). Readdition of calcium-containing buffer restored the dispersion indicating that the signal transduction by the covalent hormone-receptor complexes across the plasma membrane is also dependent on calcium.[47-49] This second site for calcium function appears to be between the receptor and the adenylate cyclase because forskolin also stimulated pigment dispersion in the absence of calcium.[50]

Extracellular magnesium did not influence MSH receptor labeling or signal transduction in *Xenopus* melanophores since experiments in the absence of Mg^{2+} and presence of EDTA plus Ca^{2+} did not differ from those with normal buffer.[48] It cannot be ruled out, however, that enough residual intracellular magnesium was present for specific functions. This also applies to calcium with respect to its apparent lack of role in adenylate cyclase stimulation by forskolin and in subsequent cyclic adenosine monophosphate (cAMP)-dependent events.

[Nle^4,D-Phe^7]α-MSH, a superagonist of α-MSH inducing long-lasting pigment dispersion similar to photoaffinity labeling,[51] has been shown to bind to the receptor also in the absence

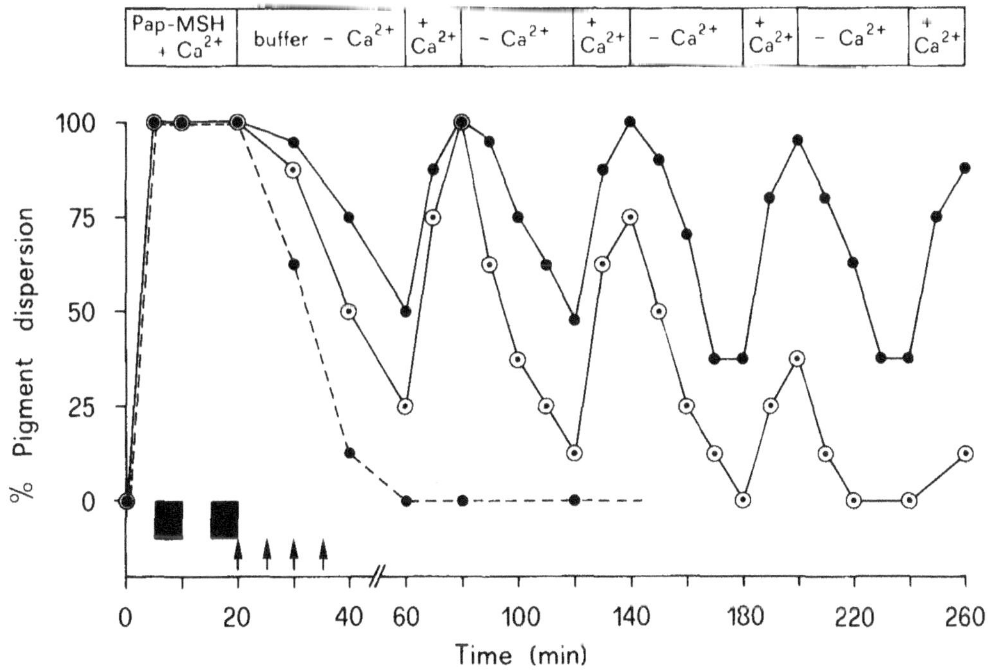

FIGURE 6. Reversible suppression of the long-lasting response in the absence of calcium.[53] Skin pieces were covalently labeled with [Pap[13]]α-MSH (●) or noncovalently with the superagonist [Nle[4],D-Phe[7]]α-MSH (⊙), then washed in calcium-free buffer, transferred back to calcium-containing buffer, and this procedure continued three times. The noncovalently attached superagonist is slowly released from the receptor in calcium-free buffer. Non-irradiated [Pap[13]]α-MSH control (--●--) is readily released.

of calcium.[52] This means that calcium is not a necessary prerequisite for MSH-receptor binding. A comparison of the long-lasting response of *Anolis* melanophores induced by photoreactive [Pap[13]]α MSH with that elicited by [Nle[4],D-Phe[7]]α-MSH demonstrated that repetitive exposure of the skins to calcium-free buffer removed the latter, but not the former (Figure 6).[53] It appears, therefore, that the superagonist did indeed bind to the receptor in calcium-free medium when the proportion of free hormone was in a large excess relative to the receptors. When, however, the ratio approached unity, the superagonist was less tightly bound in calcium-free buffer than the covalently bound photoreactive peptide. Nevertheless, it is interesting to note that the change in configuration at position 7 reduces the calcium dependence of MSH-receptor binding, but not of receptor-adenylate cyclase coupling.

C. Effect of Catecholamines

Catecholamines are well known to inhibit pigment dispersion by α-MSH in *Anolis* melanophores. Whether this inhibitory effect proceeded solely via blocking of adenylate cyclase, as shown for other hormonal systems, or whether MSH-receptor binding was also affected had not been clarified. This question has been resolved by the photoaffinity labeling approach. It was demonstrated that noradrenaline and α_1- or α_2-antagonists did not influence or only slightly affected MSH-receptor binding.[49] The coupling of the activated peptide-receptor complex to the adenylate cyclase was, however, inhibited by α_2-agonists. As shown in Figure 7, the inhibition was reversible, and the turnover of the MSH-receptor complexes did not seem to be elevated in the presence of catecholamines. α_2-Antagonists, such as yohimbine, blocked the effect of catecholamines without influencing the long-lasting signal of the α-MSH-receptor complex. The α_1-antagonist, prazosin, did not affect the inhibitory action of noradrenaline or clonidine, indicating that the modulation of the long-lasting response proceeded exclusively via α_2-receptors.[49]

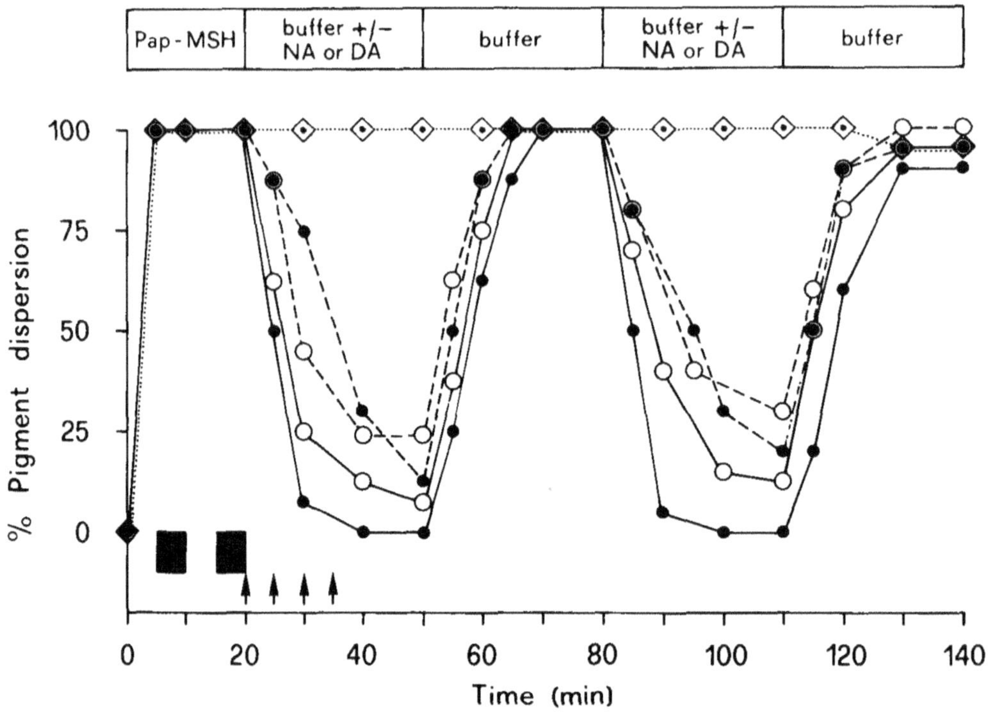

FIGURE 7. Reversible inhibition by noradrenaline and dopamine of the long-lasting stimulation of *Anolis* melanophores induced by covalent α-MSH-receptor complexes.[49] After labeling of the MSH receptors, the skins were washed in buffer containing 10^{-7} *M* (-●-) or 10^{-8} *M* (-○-) noradrenaline or 10^{-5} *M* (--●--) or 10^{-6} *M* (--○--) dopamine for 30 min. The skins were then transferred to normal buffer for 30 min and the cycle was repeated. Controls that were labeled and then maintained in normal buffer during the whole experiment remained permanently stimulated (···◇···).

Dopamine was also antagonistic, although only at a 100-fold higher concentration than noradrenaline.[49] Dopamine was first thought to act nonspecifically via α_2-adrenergic receptors, but in view of the postulated presence of D_2-receptors on mammalian follicular melanocytes,[54] it cannot be ruled out that such receptors (low-affinity type) are also present on *Anolis* melanophores.

D. Mechanism of Signal Transduction

Figure 8 shows a possible scheme of the interaction of α-MSH, α_2-receptor agonists, calcium, and forskolin in *Anolis* melanophores, as concluded from the results with photoaffinity labeling. Although the existence of the stimulatory and inhibitory coupling proteins (N_s and N_i) has not yet been unequivocally proven in this cell type, there is no reason to doubt that the principle of receptor-adenylate cyclase coupling is identical to that of the well-studied adrenergic system in, for example, erythrocytes.[55,56] A striking difference, however, is the involvement of calcium in the transduction of the α-MSH signal to the adenylate cyclase. Although in most mammalian systems the coupling proteins have been shown to contain magnesium- (or manganese-) binding sites, there have been few reports on calcium-dependent adenylate cyclase activation, such as in brain tissue or in adrenal cells.[57] Similarly, as mentioned above, *Xenopus* melanophores have also been shown to depend on extracellular calcium for α-MSH stimulation. At present, it is not yet clear whether calcium binds directly to N_s or whether a calcium-dependent regulatory protein is involved in the activation of N_s. It is interesting to note that calcium also appears to play an inhibitory role in the activation of adenylate cyclase in *Anolis* melanophores; stimulation by cholera

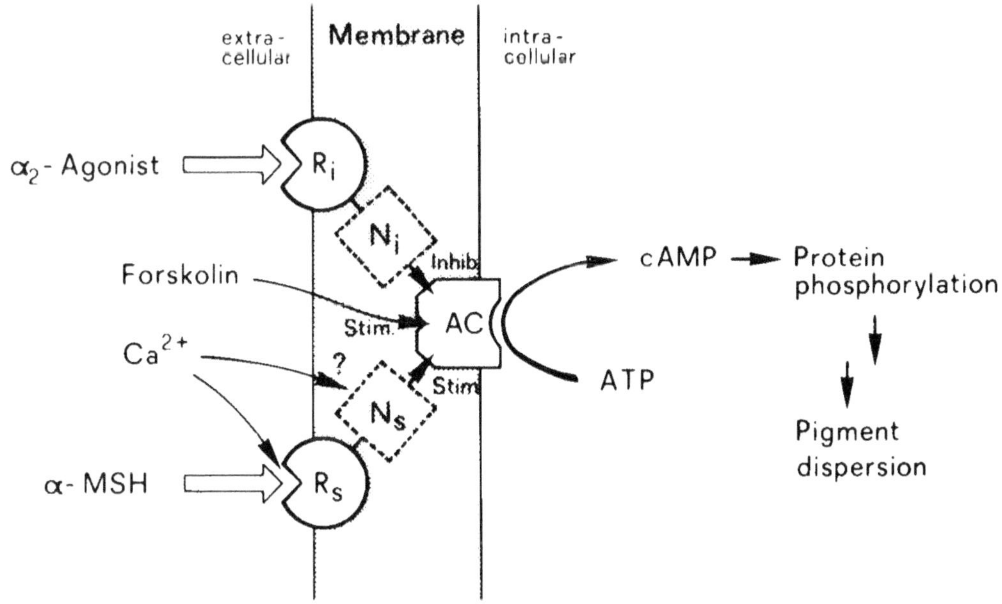

FIGURE 8. Scheme of the interaction of α-MSH, α_2-receptor agonists, and forskolin with *Anolis* melanophores.[49] R_S; MSH receptor (stimulatory); R_i; α_2-adrenergic receptor (inhibitory); N_S; stimulatory coupling protein; N_i; inhibitory coupling protein; and AC: catalytic unit of the adenylate cyclase. Forskolin stimulates AC directly, in the presence or absence of calcium, magnesium, or α_2-receptor agonists. α_2-Agonists inhibit α-MSH-induced stimulation. α-MSH requires calcium for binding and for receptor-adenylate cyclase coupling. cAMP induces protein phosphorylation which eventually leads to pigment dispersion.

toxin, which catalyzes ADP-ribosylation of the α-subunit of N_s, is extremely slow in normal buffer. In calcium-free buffer cholera toxin readily produces pigment dispersion.[53] Thus it seems that calcium may have both a stimulatory and inhibitory function in *Anolis*[49] and *Xenopus*[50] melanophores, similar to that described for Cloudman S91 mouse melanoma cells.[58]

VI. CONCLUSION

Over the past few years, biologically active monoiodinated α- and β-MSH as well as highly tritiated α-MSH derivatives have become available for the localization and physicochemical characterization of MSH receptors. The application of these compounds to new receptor-binding assays made it possible to determine MSH-receptor affinities and numbers in melanoma cells. However, the precise localization of MSH receptors on the cell surface, their fate in the cell, and their distribution and expression in melanoma tumors, are questions that have not yet been answered. It is likely that receptor autoradiography in combination with receptor-binding assays will at least resolve the problem of receptor quantification in whole tissues.

MSH derivatives containing chemical affinity labels have proved very useful for the preparation of covalent MSH-macromolecule complexes, but not for receptor labeling. In contrast, photoreactive MSH analogs have been shown to be valuable both for irreversible stimulation of functional receptors as well as for receptor identification. At present, there is, however, no consensus concerning receptor size obtained with photoaffinity and chemical cross-linking experiments. (See Addendum).

The studies on the functional characterization of MSH receptors on intact melanophores using photoaffinity labeling, clearly demonstrated that the long-lasting response is generated by permanent stimulation of the adenylate cyclase system. Although an additional second pathway of signal transduction cannot be excluded, it is very unlikely that internalization of α-MSH-receptor complexes is relevant for maintaining the response. This assumption is consistent with earlier reports that intracellularly injected α-MSH is inactive[59] and that macromolecular α-MSH-carrier complexes interact only at the cell surface. Thus, the formation of cAMP is the key regulatory element for melanosome movement in melanophores. In this respect, it has recently been shown for *Xenopus* melanophores that cAMP appears to induce pigment dispersion by phosphorylation of a 53-kdalton tubulin-like protein.[60]

Irreversible stimulation by covalent labeling of hormone receptors represents a useful approach for studying stimulus response coupling in intact cells. It is only in this model that the receptors and the adenylate cyclase are located in their different ionic environments and that an electrical potential is maintained across the plasma membrane. Thus, this system offers the possibility of studying the involvement of various factors (such as divalent ions) in receptor binding and membrane signal transduction under almost physiological conditions, even though only a limited number of cells is available.

ADDENDUM

Since submission of this chapter, considerable advances have been made in both the identification and characterization of MSH receptors on mouse and human melanoma cells. High-affinity MSH receptors are not only present on mouse B16 and S91 cells but also on most human melanoma lines.[61] MSH receptors on tumor slices can now also be identified by quantitative receptor autoradiography; this technique produces log dose-response curves which are almost identical to those obtained with isolated cells.[61] The size of the MSH receptor has recently been determined by photoaffinity labeling using biologically active, monoiodinated (Nle^4,D-Phe^7, Trp $[Naps]^9$)-α- MSH as probe. The receptor (subunit) is a 45 kda glycoprotein.[62]

ACKNOWLEDGMENTS

The work carried out in the authors' laboratories was supported by the Swiss National Science Foundation, the Schweizerische Krebsliga, and the Krebsliga beider Basel.

REFERENCES

1. **Eberle, A. N.,** Structure and chemistry of the peptide hormones of the intermediate lobe, in *Peptides of the Pars Intermedia,* Ciba Foundation Symp. No. 81, Evered, D. and Lawrenson, G., Eds., Pitman Medical, London, 1981, 13.
2. **Hruby, V. J., Wilkes, B. C., Cody, W. L., Sawyer, T. K., and Hadley, M. E.,** Melanotropins: structural, conformational and biological considerations in the development of superpotent and superprolonged analogs, in *Peptide and Protein Reviews,* Vol. 3, Hearn, M. T. W., Ed., Marcel Dekker, New York, 1984, 1.
3. **Van Nispen, J. W. and Greven, H. M.,** Structure-activity relationships of peptides derived from ACTH, β-LPH and MSH with regard to avoidance behavior in rats, *Pharmacol. Ther.,* 16, 67, 1982.
4. **Eberle, A. N., Kriwaczek, V. M., and Schulster, D.,** Corticotropin and melanotropin: organization and transmission of hormonal information, in *Perspectives in Peptide Chemistry,* Eberle, A. N., Geiger, R., and Wieland T, Eds., S. Karger, Basel, 1981, 407.
5. **Eberle, A. N. and Schwyzer, R.,** Hormone-receptor interactions. The message sequence of α-melanotropin: demonstration of two active sites, *Clin. Endocrinol.,* 5(Suppl.) 41s, 1976.
6. **Medzihradszky, K., Nikolics, K., Teplán, I., Seprödi, J., and Fittkau, S.,** Labelled polypeptides. III. Synthesis of α-melanotropin labelled with tritium on the tyrosine residue, *J. Labelled Compd.,* 10, 23, 1974.

7. **Nikolics, K., Teplán, I., and Medzihradszky, K.**, Labelled polypeptides V. Synthesis of α-melanotropin labelled with tritium on the tyrosine residue, via the dibromo intermediate, *J. Labelled Compd.*, 12, 163, 1976.
8. **Eberle, A. N. and Schwyzer, R.**, Synthesis of [D-alanine[1],4'-azido-3'-5'-ditritio-L-phenylalanine[2], norvaline[4]]-α-melanotropin as a "photoaffinity probe" for hormone-receptor interactions, *Helv. Chim. Acta*, 59, 2421, 1976.
9. **Eberle, A. N., Hübscher, W., and Schwyzer, R.**, Synthesis of radioactive α-melanotropin derivatives containing a bromoacetyl or diazoacetyl group for studies of covalent hormone-macromolecule complexes, *Helv. Chim. Acta*, 60, 2895, 1977.
10. **Eberle, A. N. and Hübscher, W.**, α-Melanotropin labelled at its tyrosine[2] residue: synthesis and biological activities of 3'-iodotyrosine[2]-, 3'-^{125}iodotyrosine[2]-, 3',5'-diiodotyrosine[2], and (3', 5'-$^3H_2$)tyrosine[2]-α-melanotropin, and of related peptides, *Helv. Chim. Acta*, 62, 2460, 1979.
11. **Buckley, D. I. and Ramachandran, J.**, Preparation of tritiated α-melanotropin with high specific radioactivity. *Intl. J. Peptide Protein Res.*, 17, 514, 1981.
12. **Wilkes, B. C., Hruby, V. J., Yamamura, H. I., Akiyama, K., de L. Castrucci, A. M., Hadley, M. E., Andrews, J. R., and Wan, Y.-P.**, Synthesis of tritium labeled Ac-[Nle[4],D-Phe[7]]-α-MSH_{4-11}-NH_2: a superpotent melanotropin with prolonged biological activity, *Life Sci.*, 34, 977, 1984.
13. **Eberle, A. N. and Zeller, A.**, Tritiation of peptides to high specific radioactivity. Synthesis and biological properties of [13-($^3H_4$)norvaline]-α-MSH and of [2,23-*bis*-($^3H_2$) tyrosine]-ACTH(1-24), *Helv. Chim. Acta*, 68, 1880, 1985.
14. **Eberle, A. N., Neubert, K., Girard, J., and Zeller, A.**, Tritiation of melanin concentrating hormone to high specific activity: synthesis of ($^3H_4$)-norvaline[3,6]-MCH, in *Peptides: Structure and Function*, Deber, C. M., Hruby, V. J., and Kopple, K. D., Eds., Pierce Chemical Company, Rockford, Ill., 1985, 667.
15. **Eberle, A. N., Neubert, K., Girard, J., and Zeller, A.**, Tritiation of peptides to high specific radioactivity. II. "Continuous flow" solid-phase synthesis and biological activity of [3,6-*bis*-($^3H_4$)norvaline]MCH, *Helv. Chim. Acta*, 1988, in press.
16. **Heward, C. B., Yang, Y. C. S., Sawyer, T. K., Bregman, M. D., Fuller B. B., Hruby, V. J., and Hadley, M. E.**, Iodination associated inactivation of β-melanocyte stimulating hormone. *Biochem. Biophys. Res. Commun.*, 88, 266, 1979.
17. **Heward, C. B., Yang, Y. C. S., Ormberg, J. F., Hadley, M. E., and Hruby, V. J.**, Effects of chloraminer T and iodination on the biological activity of melanotropin, *Hoppe-Seyler's Z. Physiol. Chem.*, 360, 1851, 1979.
18. **Heward, C. B., Kreutzfeld, K. L., Hadley, M. E., Larsen, B., Sawyer, T. K., and Hruby, V. J.**, Preparation of a radiolabeled melanotropin suitable for use as a tracer in a radioreceptor assay: [^{125}I-Tyr[2],Nle[4]]-α-melanotropin, in *Pigment Cell* 1981, Seiji, M., Ed., University of Tokyo Press, 1981, 339.
19. **Varga, J. M., Dipasquale, A., Pawelek, J., McGuire, J. S., and Lerner, A. B.**, Regulation of melanocyte-stimulating hormone action at the receptor level: discontinuous binding of hormone to synchronized mouse melanoma cells during the cell cycle, *Proc. Natl. Acad. Sci. U.S.A.*, 71, 1590, 1974.
20. **Kopp, H. G., Eberle, A. N., Vitins, P., Lichtensteiger, W., and Schwyzer, R.**, Specific antibodies against α-melanotropin for radioimmunoassay, *Eur. J. Biochem.*, 75, 417, 1977.
21. **Eberle, A. N., Jäggin, V., Siegrist, W., Scimonelli, T., and Girard, J.**, Preparation of biologically active mono-iodinated α-MSH and photoreactive α-MSH for receptor binding studies, manuscript in preparation.
22. **Lambert, D. T. and Lerner, A. B.**, Optimization of a melanotropin-receptor binding assay by reversed-phase high-performance liquid chromatography, *J. Chromatogr.*, 266, 567, 1983.
23. **Siegrist, W. and Eberle, A. N.**, *In situ* melanin assay for MSH using B16 mouse melanoma cells in culture, *Anal. Biochem.*, 159, 191, 1986.
24. **Siegrist, W., Oestreicher, M., Scimonelli, T., Stutz, S., Girard, J., Oberholzer, M., and Eberle, A. N.**, MSH receptor binding assay using cultured B16 mouse melanoma cells, manuscript in preparation.
25. **Varga, J. M., Saper, M. A., Lerner, A. B., and Fritsch, P.**, Non-random distribution of receptors for melanocyte-stimulating hormone on the surface of mouse melanoma cells, *J. Supramol. Struct.*, 4, 45, 1976.
26. **Eberle, A. N., Kriwaczek, V. M., and Schwyzer, R.**, Hormone-receptor interactions: melanotropic activities of covalent serum albumin complexes with α-melanotropin, α-melanotropin fragments, and enkephalin, *FEBS Lett.*, 80, 246, 1977.
27. **Kriwaczek, V. M., Eberle, A. N., Müller, M., and Schwyzer, R.**, Tobacco mosaic virus as a carrier for small molecules. I. The preparation and characterization of a TMV/α-melanotropin conjugate, *Helv. Chim. Acta.*, 61, 1232, 1978.
28. **DiPasquale, A., Varga, J. M., Moellmann, G., and McGuire, J.**, Synthesis of hormonally active conjugate of α-MSH, ferritin, and fluorescein, *Anal. Biochem.*, 84, 37, 1978.
29. **Bagnara, J. T. and Hadley, M. E.**, *Chromatophores and Color Change*, Prentice-Hall, Englewood Cliffs, N.J., 1973.

30. **Chaturvedi, D. N., Knittel, J. J., Hruby, V. J., de L. Castrucci, A. M., and Hadley, M. E.,** Synthesis and biological actions of highly potent and prolonged acting biotin-labeled melanotropins, *J. Med. Chem.*, 27, 1406, 1984.
31. **Eberle, A. N. and de Graan, P. N. E.,** General principles for photoaffinity labeling of peptide hormone receptors, *Methods Enzymol.*, 109, 129, 1985.
32. **Eberle, A. N., de Graan, P. N. E., and Girard, J.,** Nitrene and carbene generating photolabels for photoaffinity labelling of MSH receptors on pigment cells, in *Peptides 1984*, Ragnarsson, U., Ed., Almqvist & Wiksell, Stockholm, 1984, 391.
33. **Eberle, A. N., de Graan, P. N. E., and Hübscher, W.,** Synthesis and biological properties of *p*-azidophenylalanine^{13}-α-melanotropin, a potent photoaffinity label for MSH receptors, *Helv. Chim. Acta*, 64, 2645, 1981.
34. **Eberle, A. N., de Graan, P. N. E., and van de Veerdonk, F. C. G.,** Photoaffinity labelling of MSH receptors with [Apac-Ser^{1}]-α-MSH and [Pap^{13}]-α-MSH, in *Neuroendocrinology of Vasopressin, Corticoliberin and Opiomelanocortins*, Baertschi, A. J. and Dreifuss, J. J., Eds., Academic Press, London, 1982, 231.
35. **Eberle, A. N.,** Photoaffinity labelling of peptide hormone receptors, *J. Receptor Res.*, 3, 313, 1983.
36. **Eberle, A. N., Girard, J., de Graan, P. N. E., and van de Veerdonk, F. C. G.,** Photoreactive [Leu^{9}]-α-MSH derivatives for receptor labelling. Difference between C-terminal and N-terminal photolabel in the labelling of melanophore MSH receptors, in *Peptides 1982*, Bláha, K. and Malon, P., Eds., Walter de Gruyter, Berlin, 1983, 589.
37. **Eberle, A. N.,** Photoaffinity labelling of MSH receptors on Anolis melanophores: irradiation technique and MSH photolabels for irreversible stimulation, *J. Receptor Res.*, 4, 315, 1984.
38. **de Graan, P. N. E. and Eberle A. N.,** Irreversible stimulation of *Xenopus* melanophores by photoaffinity labelling with p-azidophenylalanine^{13}-α-melanotropin, *FEBS Lett.*, 116, 111, 1980.
39. **DiPasquale, A., McGuire, J., and Varga, J. M.,** The number of receptors for β-melanocyte stimulating hormone in Cloudman melanoma cells is increased by dibutyryl adenosine 3′:5′ cyclic monophosphate or cholera toxin, *Proc. Natl. Acad. Sci. U.S.A.*, 74, 601, 1977.
40. **Eberle, A. N., Kriwaczek, V. M., and Schwyzer, R.,** Studies on tyrosinase stimulation, binding and degradation of α-MSH interacting with non-synchronized mouse melanoma cells in culture, in *Peptides: Structure and Biological Function*, Gross, E. and Meienhofer, J., Eds., Pierce Chemical Company, Rockford Ill., 1979, 1033.
41. **Pawelek, J. M.,** Studies on the Cloudman melanoma cell line as a model for the action of MSH, *Yale J. Biol. Med.*, 58, 571, 1985.
42. **Fuller B. B. and Brooks, B. A.,** Application of percent-labeled mitoses (PLM) analysis to the investigation of melanoma cell responsiveness to MSH stimulation throughout the cell cycle, *Exp. Cell Res.*, 126, 183, 1980.
43. **Lambert, D. T., Whitcombe, P. E., Moellmann, G. E., and Lerner, A. B.,** Basic characterization of the receptor for MSH on Cloudman S91 melanoma cells, in *Pigment Cell 1985*, Bagnara, J. T., Klaus, S. N., Paul, E., and Schartl, M., Eds., University of Tokyo Press, 1985, 165.
44. **Libert, A., Arnould, R., Vercammen-Grandjean, A., and Lejeune, F.,** Binding of the α-melanocyte-stimulating hormone to human melanoma cells in culture, in *Pigment Cell 1985*, Bagnara, J. T., Klaus, S. N., Paul, E., and Schartl, M., Eds., University of Tokyo Press, 1985, 175.
45. **Bost, K. L. and Blalock, J. E.,** Molecular characterization of a corticotropin (ACTH) receptor, *Mol. Cell. Endocrinol.*, 44, 1, 1986.
46. **de Graan, P. N. E., Eberle, A. N., and van de Veerdonk, F. C. G.,** Photoaffinity labeling of MSH receptors of *Xenopus* melanophores, in *Pigment Cell 1981*, Seiji, M., Ed., University of Tokyo Press, 1981, 373.
47. **de Graan, P. N. E., Eberle, A. N., and van de Veerdonk, F. C. G.,** Photoaffinity labelling of MSH receptors reveals a dual role of calcium in melanophore stimulation, *FEBS Lett.*, 129, 113, 1981.
48. **de Graan, P. N. E., Eberle, A. N., and van de Veerdonk, F. C. G.,** Calcium sites in MSH stimulation of *Xenopus* melanophores: studies with photoreactive α-MSH, *Mol. Cell. Endocrinol.*, 26, 327, 1982.
49. **Eberle, A. N. and Girard, J.,** Photoaffinity labelling of MSH receptors on *Anolis* melanophores: effects of catecholamines, calcium and forskolin, *J. Receptor Res.*, 5, 59, 1985.
50. **de Graan, P. N. E., van de Kamp, A. J., Hup, D. R. W., Gispen, W. H., and van de Veerdonk, F. C. G.,** Calcium requirement of α-MSH action on melanophores: studies with forskolin, *J. Receptor Res.*, 4, 521, 1984.
51. **Sawyer, T. K., Sanfilippo, P. J., Hruby, V. J., Engel, M. H., Heward, C. B., Burnett, J. B., and Hadley, M. E.,** 4-Norleucine, 7-D-phenylalanine-α-melanocyte-stimulating hormone: a highly potent α-melanotropin with ultralong biological activity, *Proc. Natl. Acad. Sci. U.S.A.*, 77, 5754, 1980.
52. **Hadley, M. E., Anderson, B., Heward, C. B., Sawyer, T. K., and Hruby, V. J.,** Calcium-dependent prolonged effects on malanophores of [4-norleucine, 4-D-phenylalanine]-α-melanotropin, *Science*, 213, 1025, 1981.

53. **Eberle, A. N.,** unpublished results.
54. **Burchill, S. A. and Thody, A. J.,** Dopaminergic inhibition of tyrosinase activity in hair follicular melanocytes of the mouse, *J. Endocrinol.*, 111, 233, 1986.
55. **Ross, E. M. and Gilman, A. G.,** Biochemical properties of hormone-sensitive adenylate cyclase, *Ann. Rev. Biochem.*, 49, 533, 1980.
56. **Levitzki, A.,** Receptor to effector coupling in the receptor-dependent adenylate cyclase system, *J. Receptor Res.*, 4, 399, 1984.
57. **Bradham, L. S. and Cheung, W. Y.,** in *Calcium and Cell Function*, Vol. 1, *Calmodulin*, Cheung, W. Y., Ed., Academic Press, New York, 1980, 109.
58. **Hadley, M. E., Heward, C. B., Hruby, V. J., Sawyer, T. K., and Yang, Y. C. S.,** Biological actions of melanocyte-stimulating hormone, in *Peptides of the Pars Intermedia*, Ciba Foundation Symp. No. 81, Evered, D. and Lawrenson, G., Eds., Pitman Medical, London, 1981, 244.
59. **Horowitz, J. M., Mikukis, G. M., and Longshore, M. A.,** The response of single melanophores to extracellular and intracellular iontophoretic injection of melanocyte-stimulating hormone, *Endocrinology*, 106, 770, 1980.
60. **de Graan, P. N. E., Oestreicher, A. B., Zwiers, H., Gispen, W. H., and van de Veerdonk, F. C. G.,** Characterization of α-MSH-induced changes in the phosphorylation of a 53 kDa protein in Xenopus melanophores, *Mol. Cell. Endocrinol.*, 42, 127, 1985.
61. **Eberle, A. N.,** *The Melanotropins: Chemistry, Physiology and Mechanism of Action*, S. Karger, Basel, 1988.
62. **Scimonelli, T. and Eberle, A. N.,** Photoaffinity labelling of melanoma cell MSH receptors. *FEBS Lett.*, 226, 134, 1987.

Chapter 9

MELANOTROPIC PEPTIDES: BIOMEDICAL APPLICATIONS

Dhirendra N. Chaturvedi and Mac E. Hadley

TABLE OF CONTENTS

I. Introduction 130
A. Melanin Pigmentation of the Skin 130

II. Melanotropins and Pigmentary Disorders of the Skin 130
A. Malignant Melanoma 131
B. Hyperpigmentary Disorders 131
C. Hypomelanotic Disorders 131

III. Melanotropin Analog Possessing Unique Characteristics 132
A. Superpotent Melanotropins 132
1. [Nle^4,D-Phe^7]-Substituted Melanotropins 132
2. Cyclic Melanotropins 133
B. Melanotropins that Exhibit Prolonged Activity 133
C. Melanotropins Resistant to Enzymatic Inactivation 133

IV. Melanotropins as Diagnostic Probes 134
A. Radiolabeled Melanotropins 134
B. Biotin Melanotropin Conjugates 135
C. Fluorescein Melanotropin Conjugate 135

V. Melanotropins for Controlled Drug Delivery 135
A. The "Magic Bullet" Concept 136
B. Melanotropins as Transport Vehicles 137
1. Melanotropin Drug Conjugates 138
2. Melanotropin Toxin Conjugates 138
C. Prodrug Approach 138
D. Increased Efficacy of Transport and Uptake 139
E. Strategies of Synthesis 139
1. Carbodiimide Method 140
2. Mixed Anhydride Method 140
3. Azide Method 141
4. Active Ester Method 141
5. Oxidation-Reduction Method 141
6. Bifunctional Cross-Linking Method 142

VI. Transdermal (Topical) Delivery of Melanotropins 142

VII. Summary and Perspectives 143

Acknowledgments 145

References 145

INTRODUCTION

As has been discussed by other contributors to these volumes, melanotropic peptides may regulate a diverse number of physiological functions.[1] Besides their important role in the control of integumental melanocyte function,[2] it is likely that one or more melanotropins function within the brain as neurohormones, either as neuroregulators and/or as neurotransmitters.[3] In this chapter, we only discuss the possible biomedical applications that melanotropic peptides might serve in biology and medicine.

We have synthesized and biologically characterized a large number of α-melanotropin (α-melanocyte-stimulating hormone [α-MSH]) analogs.[4] Several of these peptides are superpotent agonists (10 to 1000 times more active than the native hormone, α-MSH) which exhibit ultraprolonged (residual) biological activity, and, in addition, are resistant to inactivation by proteolytic enzymes.[5-7] The residual melanotropic activity of these peptides may result from irreversible binding to melanocyte receptors or to some component of the pigment cell plasma membrane transduction mechanism. These melanotropins should be useful as probes in the study of normal pigment cells as well as abnormal (melanoma) cells. Conjugated to certain ligands or radiolabeled isotopes, these melanotropins might be used as proverbial "magic bullets" to unravel questions related to melanoma cell functions, such as tyrosinase activation, cellular proliferation, and the nature of the melanotropin receptor. It is hoped that the strategy of receptor targeting by melanotropins will make the use of existing antitumor drugs more effective and might, possibly, revolutionize this aspect of melanoma cancer therapy.

A. Melanin Pigmentation of the Skin

In mammals and in many other vertebrates, melanin-containing pigment cells are found in the basal layer of the epidermis. These melanocytes utilize the precursor amino acid, tyrosine, and, through a number of biochemical events controlled mainly by the enzyme, tyrosinase, produce the colored polymer, melanin. Melanin is incorporated into a subcellular structure referred to as a premelanosome. When this organelle becomes fully melanized, it is known as a melanin granule or melanosome. Melanotropins enhance melanin production within melanocytes, and the melanosomes produced migrate into the dendritic processes of the cells where they are released into the surrounding cells of the epidermis.[1,8] α-Melanotropin is the physiologically relevant hormone controlling melanin pigmentation of the skin in most vertebrates. This systemic hormone is derived from the pars intermedia of the pituitary gland. Although adult humans do not possess a pars intermedia, there is ample evidence that a melanotropin (either β-MSH or adrenocorticotropin [ACTH]) increases melanin pigmentation of the skin under certain pathological conditions, such as Addison's disease. In addition, injections of α-MSH into humans increases melanin pigmentation of the skin.[9,10] These facts suggest that, by definition, human epidermal melanocytes possess melanotropin receptors. Therefore, one could speculate that this recognition event can be utilized to direct information to the human melanocyte. Human melanocytes have, indeed, been shown to respond to agents that mimic the actions of MSH in vitro,[11] and human melanoma melanotropin receptors have been claimed to be demonstrated by the use of a radiolabeled α-MSH.[12] The effects of MSH on several mouse melanomas in culture further attest to the presence of melanotropin receptors on these cancer cells.[13-19] In this chapter we will discuss the ways in which melanotropin analogs might be used in the treatment of pigmentary disorders.

II. MELANOTROPINS AND PIGMENTARY DISORDERS OF THE SKIN

Defects in synthesis or overstimulation of the melanin synthetic pathway can lead to either hypomelanotic or hyperpigmentary disorders of the skin, and mutational changes in the melanocyte can lead to melanoma. Melanotropins are involved in the etiology of several of these disorders, but they might also be used to alleviate some of these pigmentary problems.

A. Malignant Melanoma

The human skin is populated by pigment cells, melanocytes, whose function is to produce melanin. Melanin provides a sunscreen to protect the skin from the harmful (cancer-producing) rays of the sun. Sun tanning is, in fact, an adaptive response in which these pigment cells produce even more melanin to protect the skin under conditions of continued solar exposure. Unfortunately, these pigment cells are themselves vulnerable to the mutagenic (carcinogenic) actions of solar (ultraviolet) radiation. Melanocytes are converted into cancer cells, specifically, melanoma cells. Melanoma cells proliferate to form tumors in the skin, but then melanoma cells almost invariably metastasize to other parts of the body. Early surgical removal is important, since chemotherapeutic treatment strategies are presently ineffective in eradicating the cancer.

Melanoma is notorious for its difficulty of early detection and refractoriness to chemotherapy. Worldwide statistics indicate that the incidence of melanoma is increasing.[20] The cutaneous melanomas are a group of malignancies for which there is a premium on diagnosis at the *in situ* and early invasive phases. There is a definate association between progressive depth of invasion and a worse prognosis. Clearly there is a mandate to (1) provide a method for early diagnosis, and, if metastasis has occurred to (2) provide a means for localizing the disseminated lesion as well as (3) provide a method of directing chemotherapeutic agents for eradication of the melanoma; it is also paramount to (4) understand more clearly the cellular physiology of the melanoma cell so that the above objectives can be approached with some realistic hope of success. We have developed unique melanotropin analogs that might prove to be clinically useful in the detection and eradication of primary and secondary melanotic tumors.

B. Hyperpigmentary Disorders

In humans, the amount of melanin present in the epidermis is probably regulated by the genetic background of the individual. In other words, the melanogenic activity of the integumental melanocytes is probably not under hormonal regulation. Nevertheless, the excessive release of pituitary melanotropins may result in an enhancement of melanin production within the skin for several reasons. In primary Addison's disease there is a defect in cortisol production by the adrenal cortices.[21] This results in a failure of the normal feedback inhibition by the steroid hormone on pituitary ACTH (corticotropin) secretion. The corticotrophs of the adenohypophysis, therefore, secrete excessive amounts of ACTH. Since ACTH shares a similarity in primary structure to α-MSH (Figure 1), this results in an overstimulation of melanin production by epidermal melanocytes. Hyperpigmentation is, in fact, a so-called "cardinal symptom" of Addison's disease.

In Cushing's disease of secondary origin there is an overproduction of pituitary ACTH secretion usually due to an adenoma of the corticotrophs. Again, the overstimulation of melanocytes by excessive levels of ACTH may lead to hyperpigmentation of the skin.[22]

C. Hypomelanotic Disorders

These pigmentary problems arise from an underproduction of melanin by melanocytes. If the condition is present at birth, i.e., congenital, then there is probably nothing that can be done to alleviate the problem since melanocytes are absent from the depigmented areas of the skin. This hypomelanotic disorder, referred to as piebaldism, differs from vitiligo which is defined as a progressive loss of pigmentation of the skin.[23] The etiology of vitiligo is unknown, but several hypotheses are current.[24] Chemical vitiligo may result from a destruction of melanocytes due to continued contact of the skin with certain classes of chemicals.

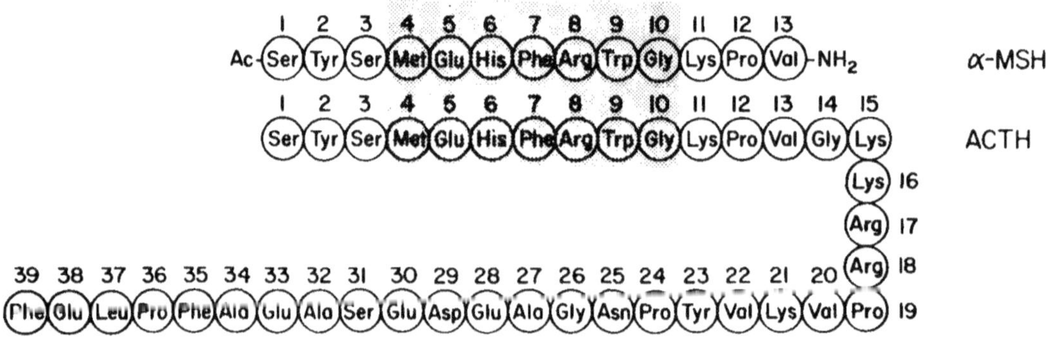

FIGURE 1. Primary structures of α-MSH and ACTH. Note the similarity in primary structure of the first 13 amino acids of ACTH with that of α-MSH. Most importantly both share a similar so-called "active site" ("message sequence").

There is some hope that the process of vitiligo might be reversed since the melanocytes are still present in the skin, at least initially. There is the possibility that repigmentation of hypomelanotic skin areas might be achieved by the use of certain highly potent melanotropin analogs as will be discussed later.

III. MELANOTROPIN ANALOGS POSSESSING UNIQUE CHARACTERISTICS

We have synthesized both natural melanotropins, α-MSH and β-MSH,[25] as well as many α-MSH analogs.[26-30] We have systematically determined the essential features of α-MSH that are necessary for biological activity. We have determined that the N-terminal and C-terminal tripeptide sequences of α-MSH are not required for melanotropin action, and we have even prepared Ac-α-MSH_{4-10}-NH_2 analogs that are more potent than the tridecapeptide sequence of the native hormone.[17,26] Our present results reveal that the minimal sequence of the native hormone for biological activity is the 6-9 sequence of α-MSH. This has allowed us to alter the hormone in a number of ways to structurally tailor the peptide for use in our studies.

In an effort to develop α-MSH analogs which possess a variety of specific and desirable biological properties, we determined the structural, stereochemical, and conformational properties associated with such biological properties.[31] The results of these investigations led us to develop melanotropins that are superpotent, some of which, in addition, are nonbiodegradable, and exhibit ultraprolonged biological activity as is discussed below.

A. Superpotent Melanotropins

We have synthesized several classes of superpotent melanotropins. Although some of these melanotropins also have prolonged action the characteristic of superpotency is related to the intrinsic structural components of peptides since some superpotent melanotropins do not exhibit prolonged activity and some prolonged-acting melanotropins are rather inactive.[4,28]

1. [Nle^4,D-Phe^7]-Substituted Melanotropins

We have synthesized a large number of [Nle^4,D-Phe^7]-substituted analogs of α-MSH. In most cases these analogs are superpotent, exhibit prolonged activity (as discussed below) and are resistant to inactivation by proteolytic enzymes. The analog [Nle^4,D-Phe^7]α-MSH was 10 to 1000 times more active than α-MSH depending upon the particular assay employed. We subsequently synthesized a number of [Nle^4,D-Phe^7]-substituted fragment analogs of α-MSH.[26] Ac-[Nle^4,D-Phe^7]α-$MSH_{4-9}NH_2$, the smallest fragment that still contain the so-called active site of α-MSH, also proved to possess the qualities of superpotency, prolonged activity, and resistance to enzymic inactivation.[33] These minimal fragment analogs may prove to be particularly useful for delivery across the skin or across a variety of mucosal epithelia.

2. Cyclic Melanotropins

We recently proposed that a β-turn or other peptide chain-reversal conformation within the central active site (His-Phe-Arg-Trp) of α-MSH may be important in the biologically active conformation of the hormone.[4,32] The cyclic disulfide tridecapeptide analog, [Cys^4,Cys^{10}]α-MSH, was then synthesized in order to evaluate the effect of covalently stabilizing a reverse turn conformation in α-MSH (see Chapter 6).[29] This peptide proved to be superpotent in stimulating melanosome dispersion within melanocytes as determined by the in vitro frog skin bioassay. Other [Cys^4,Cys^{10}]-substituted fragment analogs[34] as well as [Cys^4,D-Phe^7,Cys^{10}]-substituted analogs of α-MSH were synthesized.[35] Many of these analogs were superpotent in several bioassays. Certain [Cys^4,D-Phe^7,Cys^{10}]-substituted analogs also exhibited prolonged melanotropic activity.[35] Many of the cyclic melanotropins were resistant to serum and brain enzymes and to purified proteolytic enzymes.

B. Melanotropins that Exhibit Prolonged Activity

[Nle^4,D-Phe^7]α-MSH causes an ultraprolonged in vitro (hours)[36] and in vivo (weeks)[37] dispersion of melanin granules (melanosomes) within reptilian and amphibian integumental melanophores. The skins can be made to alternately lighten and darken as they are transferred to Ringer solution in the presence or absence of Ca^{2+}.[37] These results suggest that the analog is irreversibly bound to melanocyte melanotropin receptors or to some component of the transduction process between receptor and adenylate cyclase. If true, then it might be possible to deliver diagnostic ligands or anticancer drugs conjugated to melanotropins to melanoma cells.

We have also demonstrated that certain melanotropin analogs cause a prolonged activation of melanoma cell enzyme activation after only a short period of exposure to the cells (see Abdel-Malek, Volume II, Chapter 3). The residual (prolonged) effects of [Nle^4,D-Phe^7]-substituted melanotropin analogs were evident for days following initial contact, and subsequent removal from the culture medium. Analog-induced residual tyrosinase stimulation persisted for several days longer than that induced by α-MSH, despite continuous cell division almost every 24 hr.[15,16,19]

After an exposure of S91 melanoma cells to [Nle^4,D-Phe^7]-substituted melanotropins for only 4 hr or less (data not shown), genomic mechanisms involving both messenger RNA synthesis and protein synthesis that lead to increased tyrosinase activity are turned on for at least 6 days subsequent to removal of the hormone and in spite of continuing cell division. The interpretation of our results has important implications for cancer research and studies of cellular physiology.

C. Melanotropins Resistant to Enzymatic Inactivation

An important attribute of our superpotent and ultraprolonged acting melanotropins would be their stability in vivo or under in vitro culture conditions. α-MSH analog have, therefore, been synthesized and studied relative to their resistance to inactivation by sera and purified enzymes (α-chymotrypsin and trypsin) (see Volume I, Chapter 10). When incubated in serum, the melanotropic activity of β-MSH was lost after 4-hr incubation, while α-MSH retained about one half of its original biological activity during this same time period.[38] Both α- and β-MSH were rapidly inactivated by α-chymotrypsin and trypsin.[5,6] [Nle^4,D-Phe^7]-substituted analogs, on the other hand, were shown to be totally resistant to biological inactivation by rat serum, α-chymotrypsin, or trypsin. The conformationally restricted melanotropin $[\overline{Cys^4,Cys^{10}}]\alpha$-MSH was also totally resistant to inactivation by sera and was quite resistant to both α-chymotrypsin and trypsin.[34]

IV. MELANOTROPINS AS DIAGNOSTIC PROBES

The impressive progress of cellular and molecular biology over the last decade, supported by parallel advances in instrumentation and analytical methodologies, has generated important insights into the pathogenesis of certain disease states that are now being translated into improved clinical diagnosis and treatment. Flow cytometric, microscopic, and biochemical methods can confidently be expected to play an increasing role in a number of clinical diagnostic specialities.[39,40]

Synthetic α-MSH has been reported to bind to the plasma membrane of the HM6A human melanoma line.[12] As has been described above, the melanotropins we have developed possess prolonged biological activities and are probably irreversibly bound to melanoma cells. Since these peptides are resistant to degradation by serum enzymes, when conjugated to an appropriate ligand these melanotropins might prove useful in the diagnosis of melanoma. If radioiodinated-, biotin-, or fluorescein-labeled melanotropins also bind somewhat permanently to melanoma cells in vivo, then these melanotropins might become candidates for use in the diagnosis of human melanoma. Theoretically, any biopsy taken from a patient could be processed histologically with these melanotropins to diagnose melanoma. With a fluorescent melanotropin the histologist might even be able to detect an early invasive lesion under conditions when the malignant cells cannot be visualized by any other method. The labeled analogs might also be used as proverbial ''magic bullets'' to unravel questions related to melanoma cell functions, such as tyrosinase activation, cellular proliferation, and the nature of the melanotropin receptor. A similar approach has been used in many areas of contemporary biomedical research and for the diagnosis and staging of clinical disease.[41-43]

A. Radiolabeled Melanotropins

Radiolabeled (tritium, ^3H; iodine, ^{125}I, and ^{131}I) hormones have been utilized by many investigators to determine number and distribution of cellular hormone receptors.[44,45] α-MSH has been radioiodinated and reported to bind to cultured S91 melanoma cells.[46,47] We previously reported that α-MSH and β-MSH are inactivated by the methods generally used to radioiodinate peptide hormones.[48-50] The inactivation of the melanotropins was due to oxidation of methionine at position 4 or 7, respectively, of these peptides. We, therefore, substituted Nle at these positions and synthesized $[Nle^4]\alpha$-MSH and $[Nle^7]\beta$-MSH which were found to be resistant to the oxidation methods used to iodinate the peptides and were also more potent than the native hormones. Some iodination methods (e.g., chloramine-T) still inactivate melanotropins, but we successfully iodinated $[Nle^4]\alpha$-MSH by the iodogen method.[48] Others have also reported the successful iodination of α-MSH[51] and β-MSH.[52,53] Nevertheless, not all iodinations are successful, apparently due to interaction of iodine with amino acid residues at other sites within the melanotropin. It is hoped that further suitable substitutions within melanotropins (e.g., substitutions for His^6, Arg^8, or Trp^9) that do not lead to loss of potency may eliminate the problem of radioinactivation and provide melanotropins even more useful to the research community.

We reported the successful synthesis of a tritium-labeled melanotropin.[54] Ac-$[Nle^4$,D-$Phe^7]\alpha$-MSH_{4-11}-NH_2, an octapeptide, was a superpotent agonist of frog and lizard skin melanocytes and mouse S91 (Cloudman) melanoma cells. This melanotropin possessed ultraprolonged activity on melanocytes, both in vitro and in vivo, and the peptide was resistant to inactivation by serum and purified proteolytic enzymes. The tritium-labeled congener was prepared by direct incorporation of $[^3H]$-labeled norleucine into the peptide. The melanotropic activity of the labeled peptide, as expected, was identical to the unlabeled analog.[54]

The half-life of labeled α-MSH in man has been reported to be extremely short, 4.8 min for iodinated α-MSH and 25 min for tritiated α-MSH.[55-56] In principle, the behavior of the tritiated peptide should correspond most closely to that of the unlabeled hormone. In these studies, however, one cannot be sure that the peptide measured was the biologically active form of the molecule or a degradation product, since only radioactivity was measured and no effort was made to determine the structure of the compound containing the radioactivity. In any case, since the Ac-[(^{3}H)Nle^4,D-Phe^7]α-MSH$_{4-11}$-NH$_2$ that we have prepared was stable in serum and proteolytic enzymes for a long period of time,[54] it should provide a useful probe for the cellular localization and characterization of melanotropin receptors.

B. Biotin Melanotropin Conjugates

Because of the remarkably strong affinity of avidin and biotin (association constant of 10^{-15} *M*), biotin conjugated to proteins has been used in a number of histological investigations.[57-60] Thus, if biotin is attached through its carboxyl group to a molecule of interest, that molecule can be visualized in the electron microscope because of its electron density. Similarly, avidin-rhodamine complexes when bound to biotin can be used by fluorescence microscopy to visualize specific cellular components.[61-63] Such a fluorescent complex has been prepared and studied on mouse melanoma cells in culture. The complex was said to bind to melanoma cells, and it was reported that the fluorescent label could be visualized microscopically.[64]

We have synthesized biocytin derivatives of [Nle^4,D-Phe^7]α-MSH by solid-phase techniques.[65] The coupling of biotin and 12-aminododecanoic acid was achieved through their N-hydroxysuccinimido esters. These melanotropins (Figure 2) possessed almost identical actions to the parent analog, [Nle^4,D-Phe^7]α-MSH, as determined by several melanocyte bioassays. Both biocytin derivatives were superagonists, exhibited ultraprolonged biological activity, and were equipotent or more potent than α-MSH as determined by the melanoma tyrosinase assay. The melanotropin conjugates were also nonbiodegradable.

C. Fluorescein Melanotropin Conjugate

We have synthesized a fluorescein-labeled [Nle^4,D-Phe^7]α-MSH analog (Figure 2) using dichlorotriazinlyamino fluorescein (DTAF). Details of the synthetic methods and the bioassay results have been published.[66] Like the parent peptide, [Nle^4,D-Phe^7]α-MSH, the fluorescein analog was a superagonist, was nonbiodegradable, and possessed ultraprolonged biological activity. Most importantly, this melanotropin analog was equipotent or more potent than α-MSH on melanoma cells. This fluorescein conjugate together with the biotin-labeled melanotropins should provide useful probes for answering basic biological questions relating to the nature of melanoma melanotropin receptors. These melanotropin conjugates should also prove useful in following the mobilization and possible internalization of melanoma melanotropin receptors.

V. MELANOTROPINS FOR CONTROLLED DRUG DELIVERY

Clinical cancer chemotherapy is hindered by the lack of selective toxicity of most antitumor drugs for tumor cells relative to normal cells. A major limiting factor for continued and successful therapy is damage to all proliferating cells, including those of bone marrow, lymphoid tissue, GI and genitourinary epithelia, by presently available drugs. The need to preserve these and other vital tissues even in the face of a cancer burden reduces the maximum potential dose of cytotoxic drugs that can be effectively used. In vitro instability and degradation of the drugs before producing a desired action are yet other common problems. The main reason for these problems is that most antitumor drugs are small molecules which are transported into cells through carrier-mediated mechanisms that are not substantially

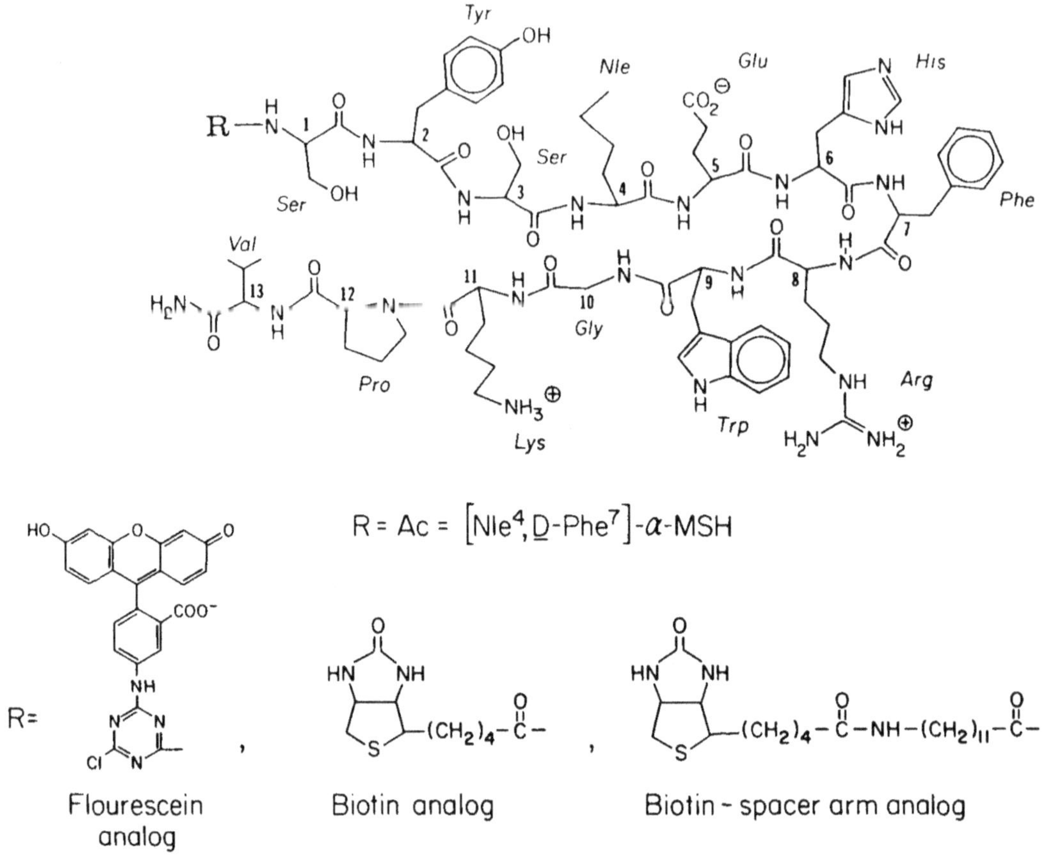

FIGURE 2. Primary structures of biocytin- and fluorescein-conjugates of [Nle^4,D-Phe^7]α-MSH.

different in normal and tumor cells, and they inhibit cell growth by interfering with biochemical pathways common to both types of cells. A possible approach for increasing the effectiveness of antitumor drugs would be to find methods of altering their distribution in the body to increase their local concentration at tumor cell sites. In this way, the selectivity of their toxicity for the tumor cells might be enhanced.

A. The "Magic Bullet" Concept

The ultimate form of controlled drug delivery would be the realization of Paul Ehrlich's "magic bullet" concept, that is, the development of drug-carrier conjugates which would deliver drugs exclusively to a particular "target" cell type. In the "magic bullet" approach, functional molecules, not necessarily target-specific, could be transported by a carrier to the site of action and released to perform their putative therapeutic tasks.

To be most therapeutically effective, a drug carrier should meet several basic requirements. First, the carrier should possess appropriate biochemical properties so that cytotoxic agents can be bound to it without adversely affecting cellular specificity of the carrier or the anticancer activity of the drug. Furthermore, the carrier should be nontoxic, although selective preferential toxicity for the target site would be beneficial. Carrier-cytotoxic agent conjugates should not bind to nontarget tissues, and the carrier as well as the transported toxic agents must not be inactivated during transit from the point of administration to the target site. Thus, maintenance of the physicochemical properties of toxic molecules by a carrier during transport followed by release at the target sites in an active form represents the ideal therapeutic situation.

Several kinds of carriers, both target-specific and nonspecific, have been used for transportation of cytotoxic drugs to effect target-cell destruction. Although carriers such as liposomes, erythrocytes, DNA, fibrinogen, and albumin, as well as other synthetic macromolecules lack tumor specificity, anticancer drugs bound to most of these agents have been reported to have increased therapeutic efficacy which can be attributed to inhibition of drug catabolism, slow release of the drug from the complex, and/or facilitation of delivery of the active drugs.[67-71] This approach to cancer therapy is based on the observation that many types of tumor cells have high endocytic activity that might allow preferential uptake of drug-macromolecule complexes.[72,73]

Specific carriers include lectins, antibodies, and hormones. Lectins have been used to probe cell surfaces because of their ability to recognize and bind to specific complex carbohydrate- components on cell surfaces. Many lectins have two or more binding sites which, in addition to forming intercellular bridges and causing agglutination also produce "caps" on target cells leading to endocytosis.[74,75] Antitumor drugs complexed to antibodies have been shown to improve cytotoxicity to malignant cells possessing antigens corresponding to the immunoglobulin carrier.[76-81] As binding to the immunological determinants of the membrane is believed to initiate endocytosis, the action of the antibody bound drug is likely to be lysosomotropic. In spite of its appeal,[79,82,83] the antibody carrier approach is associated with a number of difficulties among which the isolation of target-specific antigens and the purification from these of immunoglobulin molecules specific for the antigen are prominent.[84] Moreover, antibody mediated targeting may be of limited value in cancer therapy because most tumors do not elicit strong humoral immunity.

It is important to note that in most instances a superior life-prolonging effect was observed for antitumor drugs attached to different carriers as compared to the free drugs, and toxic effects were also often diminished.[85-90] Some of the examples are methotrexate attached to agglutinins (chlorambucil is also attached), polylysine, serum albumin, bovine fibrinogen, IgG_1 antibodies, and monoclonal antibodies. Other drugs such as mitomycin C given as a dextran conjugate, also showed increased antitumor activity. Mitomycin C, adriamycin, and daunomycin have also been linked to immunoglobulins wherein they acted as sustained release vehicles for drug release. The potency of vinca alkaloids was remarkably increased in anti-CEA IgG_1-conjugates by the ability of the antibody to deliver the drug to tumor cells.

B. Melanotropins as Transport Vehicles

Certain peptide hormones offer a unique approach as potential drug carriers since they possess a very high degree of specificity and affinity for target cells.[91-94] Using peptide hormones as carriers one should be able to enhance the therapeutic effects of a drug by augmenting the amount and persistence of drug in the vicinity of the target cell, and at the same time to minimize the toxic effects by reducing the drug exposure to nontarget cells. Similar to the concept of hormone action, hormone-drug conjugate targeting would also involve a recognition event between the drug conjugate and specific receptors at the cell surface. The drug complex would then either be internalized and release the antitumor drug for action inside the cell[16,95,96] or it might exert its cytotoxic action by interaction at the cell surface.[97,98]

MSH or related analogs have been suggested as candidates for drug transport because of their receptor specificity for melanoma cells[12,95,96,99,100] which offers real possibilities for cell-specific targeting. Daunomycin covalently linked to α-MSH has been previously utilized in such an attempt, and it was reported that the conjugate was more toxic to melanoma cells than equivalent amounts of free daunomycin.[96] Conjugates consisting of α-MSH analog attached to tobacco mosaic virus were reported to exhibit increased potencies on both amphibian and mammalian melanocytes.[101] *N*-(2-chloroethyl)-*N*-nitrosocarbamoyl, the functional group of effective antitumor agents lomustine (CCNU), carmustine (BCNU), and chlorozotocin, was utilized to make derivatives of α-MSH and fragments.[102] Another such example is a biologically active conjugate of $ACTH_{1-24}$ (structurally somewhat similar to α-MSH) with daunamycin[103] where the ED_{50} value of the conjugate was almost identical to that of $ACTH_{1-24}$. We have successfully conjugated biotin to an α-MSH analog without any loss of biological activity as discussed above (Figure 2).

1. Melanotropin Drug Conjugates

As has been already discussed, we have synthesized melanotropins that possess a number of important attributes: (1) they are the most potent of any melanotropins synthesized to date, (2) they are nonbiodegradable, (3) they exhibit extremely prolonged activity on melanocytes of a number of vertebrate systems, and (4) partial structures of α-MSH have been synthesized which are also more active than the native hormone. Thus, these peptides should provide unique substrates for the preparation of melanotropin-cytotoxic agent conjugates that could prove to be of importance in the chemotherapy of melanoma. The distribution of a melanotropin-linked cytotoxic agent will largely be dictated by the properties of the melanotropin, and accordingly, will differ from the distribution of the free agent given by the same route. As a result, melanotropin analogs would serve as a means of directing cytotoxic agents to melanotropin receptors which are localized to melanoma cells. This targeting of cytotoxic agents to melanoma, relative to tissues in which the cytotoxic effects are undesirable, should result in an improved ratio of beneficial to toxic effects of the cytotoxic agent which means that an improvement in the therapeutic index will be achieved.

2. Melanotropin Toxin Conjugates

The rationale to use toxins as the cytotoxic component of carrier molecules was provoked by their exquisite potency. The toxins bind via a recognition site on the β-chain to receptors on the cell surface, and the α-chain then penetrates (or is translocated across) the cell membrane into the cytosol where it terminates protein synthesis. Two main approaches could be adopted for the synthesis of selective cytotoxic agents from melanotropins and toxins. The first is to link the intact toxin to the melanotropin molecule and to rely upon the specific binding properties of the melanotropin to override the nonspecific binding properties of the toxin so that the net interaction is with the target melanoma cells. Perhaps the more reasonable approach is to link the isolated α-chain derived from a toxin directly to the melanotropin molecule. This maneuver would generate conjugates capable only of binding to cells expressing the receptors for melanotropins. The diphtheria toxins, abrin and ricin, have emerged as more commonly used toxins for this kind of targeting approach.[104,105]

C. Prodrug Approach

The existence of enzyme secretion by tumor cells and the presence of proteases on the tumor cell surface have been described in several experimental tumor systems. Some of these hydrolytic enzymes are, in part, selective for certain tumors, e.g., γ-glutamyltransferase, plasminogen activator, and cathepsin B. The enzyme-specific peptide derivatization of cytotoxic agents could render the drug molecule innocuous, until it is in the vicinity of the tumor or inside the tumor cell. The carrier would then cleave away to reveal the toxic warhead for its putative action at or within the target cells. These prodrugs might, therefore, be activated more or less selectively in the extracellular space of tumor tissues which have been shown to contain a high level of lysosomal enzymes and acid proteases.

Since B16 melanoma cells produce high levels of plasminogen activator, it should be possible to design specific substrates for plasmin by linking a specific peptide sequence, e.g., D-Val-Leu-Lys, to antitumor drugs to create inactive prodrugs that can be activated by plasmin. These analogs[106,107] could then be conjugated to melanotropin analogs to attain site-specific cleavage of the resulting conjugates at melanoma cells. The plasmin-activated prodrugs of the type described here could provide a reasonable prospect for improved therapeutic effectiveness in melanoma chemotherapy for the following reasons:

1. Since small amounts of plasminogen activator generate large amounts of plasmin, large amounts of enzymatic activity are available for drug activation.
2. The activity of plasminogen activator in normal tissues appears to be tightly regulated, and fibrinolysis is inherently a highly localized phenomenon due to fast-acting plasmin inhibitors in blood; thus, plasmin-mediated drug activation may also be highly localized.
3. Even though plasmin action is localized, plasminogen activation is extracellular; thus, the prodrugs will not have to cross cell membranes in order to be activated. Drugs with a wide variety of pharmacological properties and modes of action can be converted into plasmin-activated prodrugs in this manner.

D. Increased Efficacy of Transport and Uptake

Macromolecular polypeptides such as polylysin and polyglutamic acid have been used as carriers in their own right as well as functioning to alter the structure or charge of other carrier complexes. Polylysine possesses exceptional cell-membrane-penetrating properties. It is of interest to note that the methotrexate transport deficiency in tumor cells has been overcome in many experiments by the use of the polylysine carrier system.[108,109] Some measure of differential selectivity could be introduced to these molecules if they are conjugated to our unique melanotropins. The melanotropin-bound drug can ensure that elevated levels of the drug are held compartmentalized within a body cavity and thus provide a continuously higher level of drug delivery. The polymer-bound drug can also overcome transport blocks which certain cells possess and prevent passage of the free drug into the cell.

E. Strategies of Synthesis

Noncovalent complexes of drug and carrier have generally suffered from the disadvantage of being dissociable in vivo with the result that the free carrier no longer serves to transport the drug to the cancerous tissue. Covalent linkages are advantageous because they provide more stable conjugates, which are not easily split off before they reach target tissues. Although a higher ligand substitution of peptide by drug molecules is desirable, the number of drug molecules that may be attached to melanotropins should probably be limited to an optimum number of two or three since more extensive substitution may lead to a loss of drug-hormone conjugate activity.

The potential points for the conjugation of antitumor drugs to melanotropins are the carboxylic acid groups of the C-terminus and of the glutamic acid residue, and the amino groups of the N-terminus and of the lysine residue. The knowledge of linkages of varying types and stabilities is desirable in developing particular drug conjugates. Several appropriate techniques have become available through studies in related fields. The recent upsurge in the use of radioimmunoassay involving the production of antibodies to small molecules has also required the development of methods for binding hapten drug molecules to various carrier proteins without affecting their biological actions. Various chemical techniques used to modify and cross-link proteins for structural investigations are also potentially applicable. The choice of coupling conditions for drug conjugation also depends on the functional groups in the drugs that are capable of reacting without affecting drug activity. Methods that have already been used successfully, and the methods which appear to have special potential for linking antitumor drugs to the melanotropin analogs, are briefly outlined in the following paragraphs.

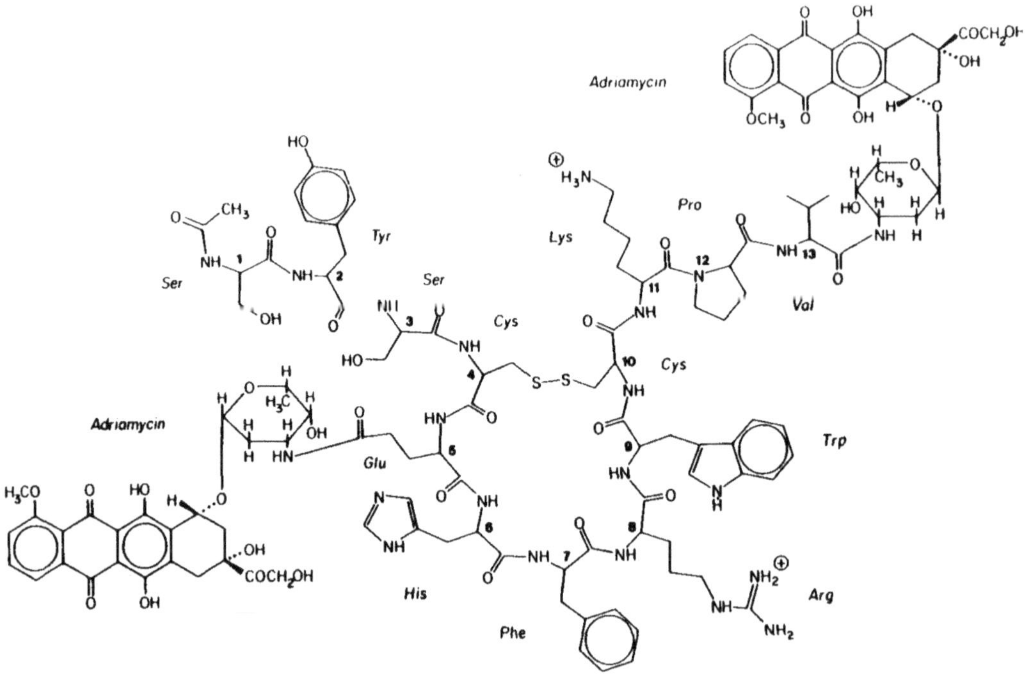

FIGURE 3. Primary structure of a putative anticancer drug (adriamycin) conjugate to a melanotropin.

1. Carbodiimide Method

This is a direct method of coupling used extensively in peptide synthesis and has been extended to the preparation of conjugates.[106] In this method, amide bonds between drug and carrier may be formed by the use of water-soluble carbodiimides, such as 1-ethyl-3-(3-dimethylaminopropyl) carbodiimide HCl and 1-cyclohexyl-3-(2-morpholinyl-[4]-ethyl)-carbodiimide metho-*p*-toluene sulfonate.[68] Methotrexate,[110,111] adriamycin, and daunomycin[112,113] have been successfully linked to the carriers by this method, and a melanotropin-adriamycin conjugate (Figure 3) could also be similarly prepared. Melphalan has been incorporated to the angiotensin II analog during peptide synthesis using dicyclohexyl carbodiimide as the coupling agent.[114] Likewise chlorambucil was conjugated to angiotensin II[115] and to bradykinin.[116] A melanotropin-melphalan conjugate (Figure 4) could similarly be prepared. It should also be noted that the coupling of methotrexate using the diazotization reaction (another method of coupling) led to numerous side-reactions in contrast to the carbodiimide procedure which showed good homogeneity with respect to the methotrexate component.[117,118]

2. Mixed Anhydride Method

This procedure, developed for peptide synthesis, was introduced for the preparation of steroid-protein conjugates. Antibodies specific for thyroxine have been obtained by using, as antigen, tetraiodothyropropionic acid coupled to protein by the mixed anhydride method. The mixed anhydride procedure has also been used for the conjugation of prostaglandins[119] and various steroids with proteins.[120] A modified mixed anhydride procedure was also used for the formation of conjugate between methotrexate and rabbit antibodies raised against mouse ovarian carcinoma. Few features of the mixed anhydride method[106,121] make it attractive in some cases, in particular, the high reaction rates at low temperatures and the generally satisfactory yields.

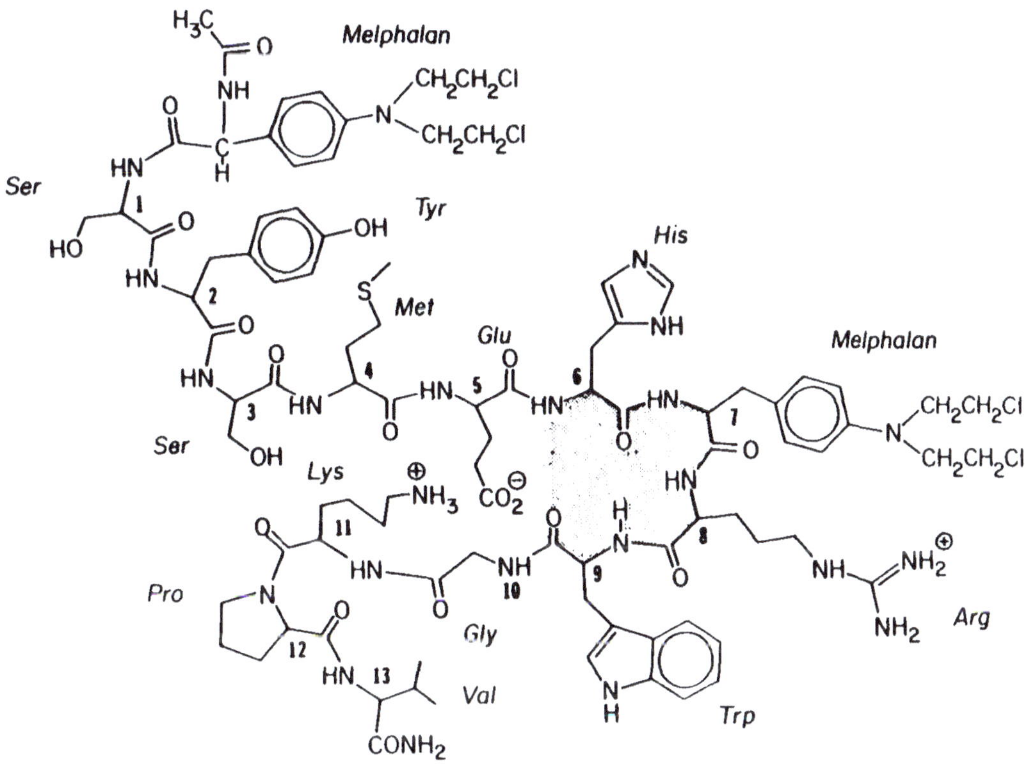

FIGURE 4. Primary structure of a putative melphalan-melanotropin drug conjugate.

3. Azide Method

An aspirin-protein conjugate was prepared by first converting aspirin to the acetylsalicylazide. The azide was coupled to rabbit serum globulin in dioxane-water solution maintained under alkaline conditions.[122] A similar procedure was used for thyroxine.[123] Although the method is generally used to achieve racemization-free couplings, it is not extensively employed for conjugation.

4. Active Ester Method

The insect juvenile hormone, epoxifarnesoic acid, was coupled to protein by the *N*-hydroxysuccinimide ester method.[124] Daunomycin was also coupled by this method to amino acids and small peptides.[125] These esters are quite stable if kept dry, but react quickly and in good yield with amino groups to form amide or peptide bonds. The method provides higher yields of more uniform products compared to the carbodiimide and mixed anhydride methods when chlorambucil is coupled to an immunoglobin.[126]

5. Oxidation-Reduction Method

The method of periodate oxidation of the drug, followed by the linking of the oxidized drug to the immunoglobulin and subsequent reduction of the product with sodium borohydride, has also been used in few instances[86,112] including conjugation of daunomycin to α-MSH.[96] The α-MSH-daunomycin conjugate prepared this way was more toxic to melanoma cells than daunomycin. The toxicity of the conjugate appeared specific for melanoma cells. The oxidation-reduction method was proved to be a method of choice in the study comparing it with carbodiimide and glutaraldehyde methods.[86]

6. Bifunctional Cross-Linking Method

A wide variety of bifunctional cross-linking reagents, i.e., homobifunctional, heterobifunctional, cleavable, and noncleavable reagents, is available and suitable for the purpose. Angiotensin has been attached by its N-terminal amino group to the amino group of a carrier by means of the bifunctional reagent xylene diisocyanate. Toluene diisocyanate has also been used in a similar manner to prepare bradykinin conjugates. Similarly, glutaraldehyde has been employed for the conjugation of ACTH, glucagon, and normetanephrine. Diphtheria toxin,[127] daunomycin, and adriamycin[83,112] were also linked to proteins following this method. It was found by comparison of carbodiimide, periodate oxidation, and glutaraldehyde methods that the drug activity was best preserved with glutaraldehyde coupling and least preserved with the carbodiimide. The use of heterobifunctional *N*-succinimidyl-3-(2-pyridyldithio) propionate (SPDP) is described by Carlsson et al.[128] In this method, essentially one introduces 2-pyridyl disulfide groups into both moieties. Treatment of one of the products with dithiothreitol (DTT) generates a thiol group. On allowing a pyridyl disulfide-substituted product to react with a thiol-substituted product, a conjugate is formed which has a central disulfide bond. This method has been used to prepare conjugates of abrin and ricin A-chains with immunoglobulins.[87,126] Diphtheria toxin A-chain[129] and a hybrid toxin[130] were similarly coupled. The SPDP method is by far the most commonly used method for conjugation work.

VI. TRANSDERMAL (TOPICAL) DELIVERY OF MELANOTROPINS

We have developed a mouse model to study cutaneous melanogenesis.[131] Injections of melanotropins in the yellow (C57BL/6JA^{Y}) mice result in a shift from phaeomelanogenesis (yellow) to eumelanogenesis (brown-black) within melanocytes of the hair follicle. The resulting change from the production of a yellow-colored melanin polymer to a black-colored melanin occurs within 24 hr post-injection of a melanotropin. Relative to the native hormone α-MSH, [Nle^4,D-Phe^7]α-MSH is at least 100 times more active in inducing the phaeomelanin-eumelanin shift. The remarkable potency of this analog in the model system is probably due to the fact that in addition to being superpotent and having prolonged action, the melanotropin is resistant to inactivation by serum enzymes.

More importantly, we have demonstrated that D-Phe^7-substituted α-melanotropin analogs can similarly enhance follicular melanogenesis following topical application to the skin of mice.[132,133] Eumelanogenesis is stimulated in follicular melanocytes as determined by both light and electron microscopic observations. The demonstration that a peptide hormone can be transdermally delivered is a unique contribution to the field of dermatology. We have also demonstrated that the peptides have not only been mobilized across the skin, but that the melanotropins have been delivered systemically as evidenced by the occurrence of eumelanogenesis in untreated distal sites. Using a human cadaver skin model, we have shown that [Nle^4,D-Phe^7]α-MSH can be delivered across the skin in vitro.[134]

These results may be of some importance for studies on melanoma. It should be possible to systemically deliver melanotropins by the transdermal route continuously over any period of time. Melanotropin-drug conjugates, if they can be similarly delivered by percutaneous absorption, should provide the optimal clinical method of drug delivery.

This concept has important implications as a method for the delivery of a drug or other agent across the skin to a site- (cell-) specific target tissue. Using a melanotropin which interacts specifically with its own cellular (melanocyte) receptors, the hormone analog may be able to deliver a cytotoxic agent or diagnostic ligand (e.g., radiolabeled peptide) to malignant melanocytes in the basal layer of the epidermis or to disseminated lesions (Figure 5). In theory, any hormone which can be delivered to the systemic circulation by percutaneous absorption might also be used to deliver a drug or other conjugate to a specific extracutaneous site. Melanotropins could prove important as a clinical method for the systemic delivery

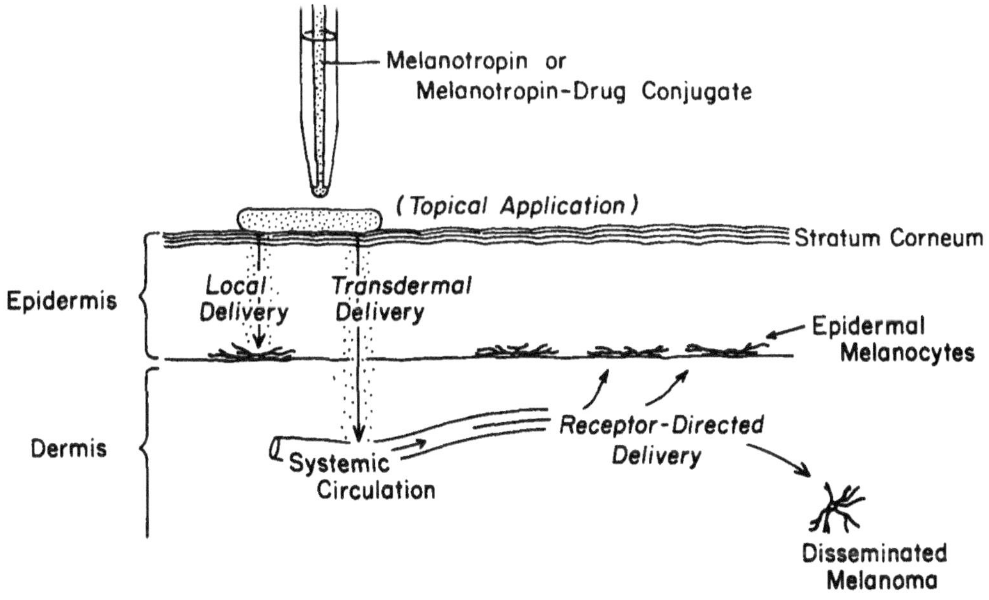

FIGURE 5. Scheme for the topical application and transdermal delivery of melanotropins and melanotropin-drug conjugates through the human skin.

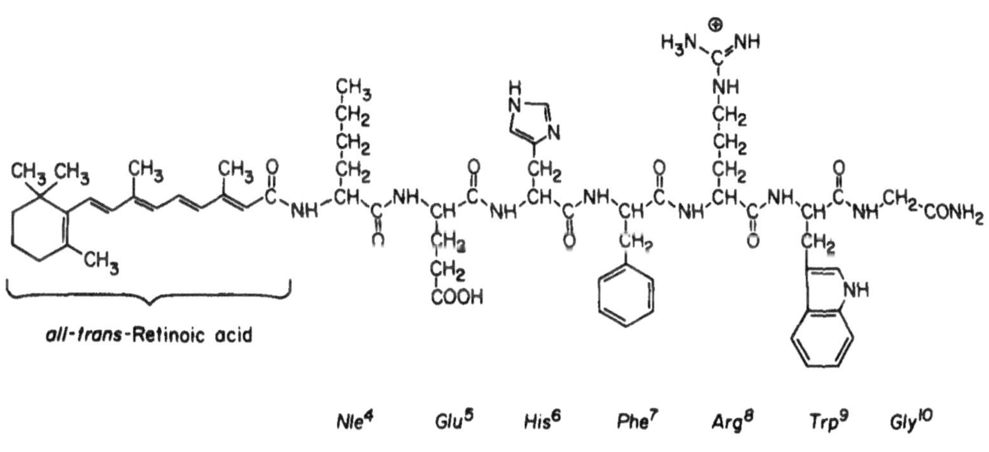

FIGURE 6. Primary structure of a putative retinoic acid-melanotropin conjugates.

(following transdermal delivery) of a melanotropin-drug conjugate. By this method it may be possible to localize by the use of a radiolabeled or a fluorescent tag malignant melanocytes that have been systemically disseminated. Melanotropin-cytotoxic drug conjugates could theoretically be used for the cellular specific destruction of malignant melanocytes. Retinoic acid has been shown to eradicate intracutaneous melanomas in mice. The use of a melanotropin-retinoic acid conjugate (Figure 6) might provide an even more selective method of melanoma eradication. The slow and continuous delivery of a melanotropin-drug conjugate by percutaneous absorption should prove a valuable contribution to the field of dermatology and clinical oncology.

VII. SUMMARY AND PERSPECTIVES

Melanotropins have been a part of biomedical research for nearly a decade, yet a comparable period of time may be required before appropriate melanotropin analogs comes into standard practice in the diagnosis and therapy of human disease. Difficulties in attaching the most effective radiolabel and its effective monitoring, together with the development of the best method to conjugate hapten drug molecules without affecting their physicochemical properties and release of the active drug at the target site, confound present investigations. However, there is much well-deserved optimism, and progress in all areas of investigation will ultimately come together to resolve most of the current problems.

Our laboratory has synthesized and biologically characterized a large number of melanotropin analogs that are commercially marketed and are being utilized by many investigators in the study of physiological problems related to melanocyte and melanoma-melanotropin receptors. The potential for the use of these and the related melanotropins in the clinic is enormous. We are understanding the biological responses of melanotropins well enough to be able to manipulate them for therapeutic purposes.

Melanoma frequently metastasizes to the liver, spleen, brain, and bone where organ-specific radionuclide or other radiographic tests are useful. The development of the effective radioiodinated melanotropins should permit safe, noninvasive, sensitive, and specific scientigraphic location of melanoma. The improved early detection of metastatic disease will allow better selection of the appropriate therapy for melanoma.

The treatment of metastatic malignant melanoma has been unsatisfactory. Dacarbazine and combinations of dacarbazine with vinblastine, bleomycin, and cisplatin have been the mainstay of therapy, although most response rates have not exceeded the range of 20 to 30%.[135] We mentioned that one reasonable approach for increasing the effectiveness of antitumor drugs would be to find methods of altering their distribution in the body to increase their local concentration at melanoma cell sites. The development of melanotropin-drug conjugates has been suggested to deliver the drug exclusively to the melanoma cells.

For a melanotropin-drug conjugate to be effective, the link between drug and carrier must remain stable in the bloodstream and withstand the action of serine hydrolases. On the other hand, unless the drug is able to act in conjugated form at the cell surface, it has to be released from the carrier after interaction of the conjugate with the target cell, and its mode of release must be such as to allow the drug to reach its biochemical target, usually situated intracellularly, and to interact effectively with it. Because the most general fate of molecules bound by surface receptors is to be internalized by endocytosis and conveyed to the lysosomes for digestion, an obvious way of ensuring appropriate release of the drug is to rely on lysosomal hydrolysis. This approach is evidently limited to drugs that are not inactivated in the lysosomes and that can reach their biochemical target from the lysosomal compartment. Drugs such as adriamycin and daunomycin will especially be suitable for this study because they have the properties of lysosome resistance and transmembrane diffusibility needed for effective action after intralysosomal release.

The melanotropin analogs developed in our laboratories and elsewhere might also be useful as pharmaceuticals for treatment of problems in learning and behavior including attention, perception, memory, and cognitive functions, and perhaps even in decreased neurostructural degeneration observed in brain aging.[136-138] Another developing area of potential promise for melanotropin drugs is the skeletal neuromuscular system in patients with pathology of motor neurons such as multiple sclerosis, progressive spinal muscular atrophy, and carpal tunnel syndrome.[139,140] α-MSH has been suggested as a very useful antipyretic, clinically.[141] The centrally administered melanotropin is much more potent in reducing fever than the widely used antipyretic acetaminophen. The molecule does not show any irreversible deleterious effects when very large doses are given to man.

It appears inevitable that further research in these areas will have a significant impact on melanotropin drug development studies for a variety of biomedical applications.

ACKNOWLEDGMENTS

This work was supported, in part, by grants from the U.S. Public Health Service (AM 17420) and the National Science Foundation (PCM 77-07051 and PCM-81-0078).

REFERENCES

1. **Hadley, M. E.,** *Endocrinology,* Prentice-Hall, Englewood Cliffs, N.J., 1984, chap. 8.
2. **Bagnara, J. T. and Hadley, M. E.,** *Chromatophores and Color Change: Comparative Physiology of Animal Pigmentation,* Prentice-Hall, Englewood Cliffs, N.J., 1972.
3. **O'Donohue, T. K. and Dorsa, D. M.,** The opiomelanotropinergic neuronal and endocrine systems, *Peptides,* 3, 353, 1982.
4. **Hruby, V. J., Wilkes, B. C., Cody, W. L., Sawyer, T. K., and Hadley, M. E.,** Melanotropins: structural, conformational and biological considerations in the development of superpotent and superprolonged analogs, in *Protein Peptide Review,* Vol. 3, Hearn, M. T. W., Ed., Marcel Dekker, New York, 1984, 1.
5. **de L. Castrucci, A. M., Hadley, M. E., Yorulmazoglu, E. I., Wilkes, B. C., Sawyer, T. K., and Hruby, V. J.,** Synthesis and studies of superpotent melanotropins resistant to enzyme degradation, in *Biological, Molecular and Clinical Aspects of Pigmentation,* Bagnara, J., Klaus, S. N., Paul, E., and Schartl, M., Eds., University of Tokyo Press, 1985, 145.
6. **de L. Castrucci, A. M., Hadley, M. E., Sawyer, T. K., and Hruby, V. J.,** Enzymological studies of melanotropins, *Comp. Biochem. Physiol.,* 78B, 519, 1984.
7. **Akiyama, K., Yamamura, H. I., Wilkes, B. C., Cody, W. L., Hruby, V. J., de L. Castrucci, A. M., and Hadley, M. E.,** Relative stability of α-melanotropin and related analogues to rat brain homogenates, *Peptides,* 5, 1191, 1984.
8. **Hadley, M. E. and Quevedo, W. C., Jr.,** Vertebrate epidermal melanin unit, *Nature (London),* 209, 1334, 1966.
9. **Lerner, A. B. and McGuire, J. S.,** Effect of alpha- and beta-melanocyte stimulating hormones on the skin color of man, *Nature (London),* 189, 176, 1961.
10. **Lerner, A. B. and McGuire, J. S.,** Melanocyte-stimulating hormone and adrenocorticotrophic hormone, their relationship to pigmentation, *N. Engl. J. Med.,* 270, 539, 1964.
11. **Kitano, Y.,** Effects of dibutyryl adenosine 3′,5′-cyclic monophosphate on human melanocytes in culture, *Acta Derm. Venereol. (Stockholm),* 56, 223, 1971.
12. **Legros, F., Coele, J., Doyen, A., Hanson, P., Van Tieghem, N., Vercammen-Grandjean, A., Fruhling, J., and Lejeune, F. J.,** α-Melanocyte-stimulating hormone binding and biological activity in a human melanoma cell line, *Cancer Res.,* 41, 1539, 1981.
13. **Pawelek, J.,** Factors regulating growth and pigmentation of melanoma cells, *J. Invest. Dermatol.,* 66, 201, 1976.
14. **Kreiner, P. W., Gold, C. J., Keirns, J. J., Brock, W. A., and Bitensky, M. W.,** MSH-sensitive adenyl cyclase in the Cloudman melanoma, *Yale J. Biol. Med.,* 46, 583, 1973.
15. **Abdel-Malek, Z. A., Hadley, M. E., Bregman, M. D., Meyskens, F. L., Jr., and Hruby, V. J.,** Actions of melanotropins on mouse melanoma cell growth *in vitro, J. Natl. Cancer Inst.,* 76, 857, 1985.
16. **Abdel-Malek, Z. A., Kreutzfeld, K. L., Hruby, V. J., Bregman, M. D., Meyskens, F. L., Jr., and Hadley, M. E.,** Melanotropin stimulation of S91 melanoma cell tyrosinase is density dependent, *In Vitro,* 22, 75, 1986.
17. **Marwan, M. M., Z. Abdel Malek, Z. A., Kreutzfeld, K. L., Hadley, M. E., Wilkes, B. C., Hruby, V. J., and de L. Castrucci, A. M.,** Stimulation of S91 melanoma tyrosinase activity by superpotent α-melanotropins, *Mol. Cell. Endocrinol.,* 44, 171, 1985.
18. **Bregman, M. D., Sawyer, T. K., Hadley, M. E., and Hruby, V. J.,** Adenosine and divalent cation effects on S91 melanoma adenylate cyclase, *Arch. Biochem. Biophys.,* 200, 1, 1980.
19. **Hadley, M. E., Abdel-Malek, Z. A., Marwan, M. M., Kreutzfeld, K. L., and Hruby, V. J.,** [Nle[4],D-Phe[7]]-α-MSH: a superpotent melanotropin that "irreversibly" activates melanoma tyrosinase, *Endocr. Res. Commun.,*
20. **Elwood, J. M., Gallagher, R. P., Hill, G. B., and Pearson, J. C. G.,** Cutaneous melanoma in relation to intermittent and constant sun exposure — the Western Canada melanoma study, *Intl. J. Cancer,* 35, 427, 1985.
21. **Mulligan, T. M. and Sorvers, J. R.,** Hyperpigmentation, vitiligo, and Addison's disease, *CUTIS,* October, 317, 1985.

22. **McArthur, R. G., Cloutier, M. D., Hayles, A. B., and Sprague, R. G.,** Cushing's disease in children, *Mayo Clin. Proc.*, 47, 318, 1972.
23. **Lerner, A. B.,** On the etiology of vitiligo and gray hair, *Am. J. Med.*, 51, 141, 1971.
24. **Lerner, A. B. and Nordlund, J. J.,** Vitiligo, what is it? Is it important?, *JAMA*, 239, 1183, 1978.
25. **Yang, Y. C. S., Hruby, V. J., Heward, C. B., and Hadley, M. E.,** Synthesis of α- and β-melanocyte stimulating hormones, *Intl. J. Pept. Protein Res.*, 15, 130, 1979.
26. **Sawyer, T. K., Hruby, V. J., Wilkes, B. C., Draelos, M. T., Hadley, M. E., and Bergsneider, M.,** Comparative biological activity of highly potent active-site analogues of α-melanotropin, *J. Med. Chem.*, 25, 1022, 1982.
27. **Sawyer, T. K., Sanfillippo, P. J., Hruby, V. J., Engel, M. H., Heward, C. B., Burnett, J. B., and Hadley, M. E.,** 4-Norleucine, 7-D-phenylalanine-α-melanocyte-stimulating hormone: a highly potent α melanotropin with ultraprolonged biological activity, *Proc. Natl. Acad. Sci. U.S.A.*, 77, 5754, 1980.
28. **Wilkes, B. C., Sawyer, T. K., Hruby, V. J., and Hadley, M. E.,** Differentiation of the structural features of melanotropins important for biological potency and prolonged activity *in vitro, Intl. J. Pept. Protein Res.*, 22, 313, 1983.
29. **Sawyer, T. K., Hruby, V. J., Darman, P. S., and Hadley, M. E.,** [Half-Cys^4,half-Cys^{10}]-α-melanocyte stimulating hormone: a cyclic α-melanotropin exhibiting superagonist biological activity, *Proc. Natl. Acad. Sci. U.S.A.*, 79, 1751, 1982.
30. **Knittel, J. J., Sawyer, T. K., Hruby, V. J., and Hadley, M. E.,** Structure-activity studies of highly potent cyclic [Cys^4, Cys^{10}]-melanotropin analogues, *J. Med. Chem.*, 28, 125, 1983.
31. **Sugg, E. E., Cody, W. L., Abdel-Malek, Z. A., Hadley, M. E., and Hruby, V. J.,** Isomeric replacements and within the 6-9 core sequence of Ac-[Nle^4]-α-MSH_{4-11}-NJ_2: a topological model for the solution conformation of α-melanotropin, *Biopolymers*, 25, 2029, 1986.
32. **Sawyer, T. K., Hruby, V. J., Hadley, M. E., and Engel, M. H.,** α-Melanocyte stimulating hormone: chemical nature and mechanism of action, *Am. Zool.*, 23, 529, 1983.
33. **Hadley, M. E., Heward, C. B., Hruby, V. J., Sawyer, T. K., and Yang, Y. C. S.,** Hormone receptors of vertebrate pigment cells, in *Pigment Cell 1981,* Seiji, M. E., Ed., University of Tokyo Press, 1981, 323.
34. **Cody, W. L., Wilkes, B. C., Muska, B. J., Hruby, V. J., de L. Castrucci, A. M., and Hadley, M. E.,** Cyclic melanotropins. V. Importance of the C-terminal tripeptide, (Lys-Pro-Val), *J. Med. Chem.*, 27, 1186, 1984.
35. **Cody, W. L., Mahoney, M., Knittel, J. J., Hruby, V. J., de L. Castrucci, A. M., and Hadley, M. E.,** Cyclic melanotropins. IX. 7-D-phenylalanine analogues of the active-site sequence, *J. Med. Chem.*, 28, 583, 1985.
36. **Hadley, M. E., Mieyr, J. H., Martin, B., de L. Castrucci, A. M., Hruby, V. J., Sawyer, T. K., and Rao, R.,** [Nle^4,D-Phe^7]-α-MSH: a melanotropin with superagonist and prolonged action on vertebrate chromatophores, *Comp. Biochem. Physiol.*, 81A, 1, 1984.
37. **Hadley, M. E., Anderson, B., Heward, C. B., Sawyer, T. K., and Hruby, V. J.,** Calcium-dependent prolonged effects on melanophores of [4-norleucine,7-D-phenylalanine] α-melanotropin, *Science*, 213, 1025, 1981.
38. **de L. Castrucci, A. M., Hadley, M. E., and Hruby, V. J.,** Melanotropin bioassays: *in vitro* and *in vivo* comparisons, *Gen. Comp. Endocrinol.*, 55, 104, 1984.
39. **Muirhead, K. A., Horan, P. K., and Poste, G.,** Flow cytometry: present and future, *Biotechnology*, 3, 337, 1985.
40. **Cheung, A. T. W., Cox, K. L., Miller, M. E., and Walsh, E. M.,** Fluorescence intravital microscopy: a new approach for *in vivo* investigations, *Biotechniques*, 4, 346, 1986.
41. **Bentley, K. L., Thompson, L. K., Klebe, R. J., and Horowitz, P. M.,** Fluorescence polarization: a general method for measuring ligand binding and membrane microviscosity, *Biotechniques*, 3, 356, 1985.
42. **DiGuiseppi, J., Inman, R., Ishihara, A., Jacobson, K., and Herman, B.,** Applications of digitized fluorescence microscopy to problems in cell biology, *Biotechniques*, 3, 394, 1985.
43. **Levine, A.,** On the histological diagnosis and prognosis of malignant melanoma, *J. Clin. Pathol.*, 33, 101, 1980.
44. **Goldenberg, D. M., Kim, E. E., and DeLand, F. H.,** Human chorionic gonadotropin radioantibodies in the radioimmunodetection of cancer and for disclosure of occult metastases, *Proc. Natl. Acad. Sci. U.S.A.*, 78, 7754, 1981.
45. **Mack, J. P., Carrel, S., Forni, M., Ritschard, J., Donath, A., and Alberto, P.,** Tumor localization of radiolabeled antibodies against carcinoembryonic antigen in patients with carcinoma, *N. Engl. J. Med.*, 303, 5, 1980.
46. **Varga, J. M., Moellmann, G., Fritsch, P., Godawska, E., and Lerner, A. B.,** Association of cell surface receptors for melanotropin with the Golgi region in mouse melanoma cells, *Proc. Natl. Acad., Sci. U.S.A.*, 73, 559, 1976.

47. **Varga, J. M., DiPasquale, A., Pawelek, J., McGuire, J. S., and Lerner, A. B.**, Regulation of melanocyte-stimulating hormone action at the receptor level: discontinuous binding of hormone to synchronized mouse melanoma cells during the cell cycle, *Proc. Natl. Acad. Sci. U. S. A.*, 72, 1590, 1974.
48. **Heward, C. B., Kreutzfeld, K. L., Hadley, M. E., Larsen, B., Sawyer, T. K., and Hruby, V. J.**, Preparation of a radiolabeled melanotropin suitable for use as a tracer in a radioreceptor assay: [^{125}I-Tyr^2,Nle^4]-α-MSH, in *Phenotypic Expression in Pigment Cells*, Seiji, M., Ed., University of Tokyo Press, 1981, 339.
49. **Heward, C. B., Yang, Y. C. S., Sawyer, T. K., Bregman, M. D., Fuller, B. B., Hruby, V. J., and Hadley, M. E.**, Iodination associated inactivation of α-melanocyte stimulating hormone, *Biochem. Biophys. Res. Commun.*, 88, 266, 1979.
50. **Heward, C. B., Yang, Y. C. S., Ormberg, J. F., Hadley, M. E., and Hruby, V. J.**, Effects of chloramine T and iodination on the biological activity of α-melanotropin, *Hoppe-Seyler's Z. Physiol. Chem.*, 360, 1851, 1979.
51. **Eberle, A. and Hubscher, W.**, α-Melanotropin labelled at its tyrosine^2 residue: synthesis and biological activities of 3^1-iodotyrosine^2, 3-^{125}iodotyrosine^2, 3′,5′-diiodotyrosine^2-, and (3′,5′-$^3H_2$) tyrosine^2-α-melanotropin, and of related peptides, *Helv. Chim. Acta*, 62, 2460, 1979.
52. **Lambert, D. T., Stachelek, C., Varga, J. M., and Lerner, A. B.**, Iodination of α-melanotropins, *J. Biol. Chem.*, 257, 8211, 1982.
53. **Lambert, D. T., Whitcombe, P. E., Moellmann, G. E., and Lerner, A. B.**, Basic characterization of the receptor for MSH on Cloudman S91 melanoma cells, in *Pigment Cell 1985*, Bagnara, J., Klaus, S. N., and Schartl, M., Eds., University of Tokyo Press, 1985, 165.
54. **Wilkes, B. C., Hruby, V. J., Yamamura, H. I., Akiyama, K., de L. Castrucci, A. M., Hadley, M. E., Andrews, J. R., and Wan, Y.-P.**, Synthesis of tritium labeled Ac-[Nle^4,D-Phe^7]-α-MSH_{4-11}-NH_2: a superpotent melanotropin with prolonged biological activity, *Life Sci.*, 34, 977, 1983.
55. **Redding, T. W., Kastin, A. J., Nikolics, K., Schally, A. V., and Coy, D., H.**, Disappearance and excretion of labeled α-MSH in man, *Pharmacol. Biochem. Behav.*, 9, 207, 1978.
56. **Marks, N., Stern, F., and Kastin, A. J.**, Biodegradation of α-MSH and derived peptides by rat brain extracts, and by rat and human serum, *Brain Res. Bull.*, 1, 591, 1976.
57. **Finn, F. M. and Hofmann, K. H.**, Synthesis of biotinyl derivatives of peptide hormones and other biological materials, *Methods Enzymol.*, 109, 418, 1985.
58. **Billingsley, M. L., Pennypacker, K. R., Hoover, C. G., and Kincaid, R. L.**, Biotinylated proteins as probes of protein structure and protein-interactions, *Biotechniques*, 5, 22, 1987.
59. **Holtzman, E., Wise, D., Wall, J., and Karlin, A.**, Electron microscopy of complexes of isolated acetylcholine receptor, biotinyl-toxin and avidin, *Proc. Natl. Acad. Sci. U.S.A.*, 79, 310, 1982.
60. **Childs, G. V., Naor, Z., Hazum, E., Tibolt, R., Westlund, K. N., and Hancock, M. B.**, Localization of biotinylated gonadotropin releasing hormone on pituitary monolayer cells with avidin-biotin-peroxidase complexes, *J. Histochem. Cytochem.*, 31, 1422, 1983.
61. **Bayer, E. A. and Wilchek, M. A.**, The use of the avidin-biotin complex as a tool in molecular biology, *Methods Biochem. Anal.*, 26, 1, 1980.
62. **Guesdon, J. L., Ternynck, T., and Avremeas, S. J.**, The use of avidin-biotin interaction in immunoenzymatic techniques, *J. Histochem. Cytochem.*, 27, 1131, 1979.
63. **Leary, J. J., Brigati, D. J., and Ward, D. C.**, Rapid and sensitive colorimetric method for visualizing biotin-labeled DNA probes hybridized to DNA or RNA immobilized to nitrocellulose: bio-blots, *Proc. Natl. Acad. Sci. U.S.A.*, 80, 4045, 1983.
64. **DiPasquale, A., Varga, J. M., Moellmann, G., and McGuire, J.**, Synthesis of a hormonally active conjugate of α-MSH, ferritin, and fluorescein, *Anal. Biochem.*, 84, 37, 1978.
65. **Chaturvedi, D. N., Knittel, J. J., Hruby, V. J., de L. Castrucci, A. M., and Hadley, M. E.**, Highly potent peptide hormone analogues: synthesis and biological actions of biotin-labeled melanotropins, *J. Med Chem.*, 27, 1406, 1984.
66. **Chaturvedi, D. N., Hruby, V. J., de L. Castrucci, A. M., and Hadley, M. E.**, Synthesis and biological actions of a fluorescein-labeled superpotent melanotropin, *J. Pharm. Sci.*, 74, 237, 1985.
67. **Rubens, R. D.**, Antibodies as carriers of anticancer agents, *Lancet*, 1, 498, 1974.
68. **Szekereke, M. and Driscoll, J. S.**, The use of macromolecules as carriers of antitumor drugs, *Eur. J. Cancer*, 13, 529, 1977.
69. **Gregoriadis, G.**, Targeting of drugs, *Nature (London)* 265, 407, 1977.
70. **Trouet, A., Masquelier, M., Baurain, R., and Deprez-de Campeneere, D.**, A covalent linkage between daunorubicin and proteins that is stable in serum and reversible by lysosomal hydrolases, as required for a lysosomotropic drug-carrier conjugate: *in vitro* and *in vivo* studies, *Proc. Natl. Acad. Sci. U.S.A.*, 79, 626, 1982.
71. **Wiley, R. G., Blessing, W. W., and Reis, D. J.**, Suicide transport: destruction of neurons by retrograde transport of ricin, abrin, and modeccin, *Science*, 21, 889, 1982.

72. **Silverstein, S. C., Steinman, R. M., and Cohn, Z. A.,** Endocytosis, *Ann. Rev. Biochem.*, 46, 669, 1977.
73. **Venter, J. C.,** Immobilized and insolubilized drugs, hormones, and neurotransmitters: properties, mechanisms of action and applications, *Pharmacol. Rev.*, 34, 153, 1982.
74. **Nicolson, G. L.,** The interactions of lectins with animal cell surfaces, *Intl. Rev. Cytol.*, 39, 90, 1974.
75. **Rapin, A. M. and Burger, M. M.,** Tumor cell surfaces: general alterations detected by agglutinins, *Adv. Cancer Res.*, 20, 1, 1974.
76. **Tai, J., Blair, A. H., and Ghose, T.,** Tumor inhibition by chlorambucil covalently linked to antitumor globulin, *Eur. J. Cancer*, 15, 1357, 1979.
77. **Rowland, G. F.,** Effective antitumor conjugates of alkylating drug and antibody using dextran as the intermediate carrier, *Eur. J. Cancer*, 13, 593, 1977.
78. **Distefana, P. S., Schweitzer, J. B., Taniuchi, M., and Johnson, E. M., Jr.,** Selective destruction of nerve growth factor receptor bearing cells *in vitro* using a hybrid toxin composed of ricin A chain and a monoclonal antibody against the nerve growth factor receptor, *J. Cell Biol.*, 101, 1107, 1985.
79. **Edwards, D. C.,** Targeting potential of antibody conjugates, *Pharmacol. Ther.*, 23, 147, 1983.
80. **Krolick, K. A., Uhr, J. W., Slavin, S., and Vietta, E. S.,** *In vivo* therapy of a murine B cell tumor (BCL), using antibody-ricin A chain immunotoxins, *J. Exp. Med.*, 155, 1797, 1982.
81. **Tsukada, Y., Hurwitz, E., Kashi, R., Sela, M., Hibi, N., Hara, A., and Hirai, H.,** Chemotherapy by intravenous administration of conjugates of daunomycin with monoclonal and conventional anti-rat α-fetoprotein antibodies, *Proc. Natl. Acad. Sci. U.S.A.*, 79, 7896, 1982.
82. **Raso, V. and Basala, M.,** Monoclonal antibodies as cell targeted carriers of covalently and noncovalently attached toxins, in *Receptor-Mediated Targeting of Drugs* , Gregoriadis, G., Poste, G., Senior, J., and Trouet, A., Eds., Plenum Press, N.Y., 1984, 119.
83. **Lee, F. H. and Hwang, K. M.,** Antibodies as specific carriers for chemotherapeutic agents, *Cancer Chemother. Pharmacol.*, 3, 17, 1979.
84. **Ghose, T. and Blair, A. H.,** Antibody-linked cytotoxic agents in the treatment of cancer: current status and future prospects, *J. Natl. Cancer Inst.*, 61, 657, 1978.
85. **Poste, G.,** Drug targeting in cancer therapy, in *Receptor-Mediated Targeting of Drugs*, Gregoriadis, G., Poste, G., Senior, J., and Trouet, A., Eds., Plenum Press, N.Y., 1984, 427.
86. **Arnon, R. and Sela, M.,** *In vitro* and *in vivo* efficacy of conjugates of daunomycin with antibodies, *Immunol. Rev.*, 62, 5, 1982.
87. **Jansen, F. K., Blythman, H. E., Carriere, D., Casellas, P., Gros, O., Gros, P., Laurent, J. C., Paolucci, F., Pau, B., Poncelet, P., Richer, G., Vidal, H., and Voisin, G. A.,** Immunotoxins: hybrid molecules combining high specificity and potent cytotoxicity, *Immunol. Rev.*, 62, 185, 1982.
88. **Raso, V.,** Antibody mediated delivery of toxic molecules to antigen bearing target cells, *Immunol. Rev.*, 62, 93, 1982.
89. **Edwards, D. C. and Thorpe, P. E.,** Targeting toxins — the retiarian approach to chemotherapy, *Trends Biochem. Sci.*, 6, 313, 1981.
90. **Bumol, T. F., Wang, Q. C., Reisfeld, R. A., and Kaplan, N. O.,** Monoclonal antibody and an antibody-toxin conjugate to a cell surface proteoglycan of melanoma cells suppress *in vivo* tumor growth, *Proc. Natl. Acad. Sci. U.S.A.*, 80, 529, 1983.
91. **Lee, A. C., Powell, J. E., Tregear, G. W., Niall, H. D., and Stevens, V. C.,** A method for preparing B-HCG COOH peptide-carrier conjugates of predictable composition, *Mol. Immunol.*, 17, 749, 1980.
92. **Nixon, D. W.,** Melanoma, in *Diagnosis and Management of Cancer*, Nixon, D. W., Ed., Addison-Wesley Publishing, California, 1982, 237.
93. **Cawley, D. B., Herschman, H. R., Gilliland, D. G., and Collier, R. J.,** Epidermal growth factor-toxin A chain conjugates: EGF-ricin A is a potent toxin while EGF-diphtheria fragment A is nontoxic, *Cell*, 22, 563, 1980.
94. **Oeltman, T. N. and Heath, E. C.,** A hybrid protein containing the toxic subunit of ricin and the cell-specific subunit of human chorionic gonadotropin. II. Biologic properties, *J. Biol. Chem.*, 254, 1028, 1979.
95. **Pawelek, J. M. and Korner, A. M.,** The biosynthesis of mammalian melanin, *Am. Sci.*, 70, 136, 1982.
96. **Varga, J. M., Asato, N., Lande, S., and Lerner, A. B.,** Melanotropin-daunomycin conjugate shows receptor-mediated cytotoxicity in cultured murine melanoma cells, *Nature (London)*, 267, 5606, 1977.
97. **Tritton, T. R. and Yee, G.,** The anticancer agent adriamycin can be actively cytotoxic without entering cells, *Science*, 217, 248, 1982.
98. **Venter, J. C.,** Immobilized and insolubilized drugs, hormones, and neurotransmitters: properties, mechanisms of action and applications, *Pharmacol. Rev.*, 34, 153, 1982.
99. **Varga, J. M. and Asato, N.,** Hormones as drug carriers, in *Targeted Drugs*, Goldberg, E., Ed., John Wiley & Sons, New York, 1983, 73.
100. **Thiers, B. H.,** Chemotherapy of malignant melanoma, *J. Am. Acad. Dermatol.*, 6, 412, 1982.
101. **Kriwaczek, V. M., Eberle, A., Muller, M., and Schwyzer, R.,** Tobacco mosaic virus as a carrier for small molecules. I. The preparation and characterization of TMV/α-melanotropin conjugate, *Helv. Chim. Acta*, 61, 1232, 1978.

102. **Vargha, H. S. and Medzihradszky, K.**, Synthesis of N-(2-chloroethyl)-*N* nitrosocarbamoyl derivatives of biologically active polypeptide hormone fragments, *Intl. J. Pept. Protein Res.*, 23, 650, 1984.
103. **Scott, D. and Ontjes, D.**, Biologically active conjugates of ACTH and cytotoxic drugs: properties of ACTH analogs containing daunorubicin, in *Peptides: Synthesis, Structure, Function*, Proc. 7th Am. Pept. Symp., Rich, D. H. and Gross, E. Eds. Pierce Chemical Company, Rockford, ILL., 1981, 817.
104. **Bacha, P., Murphy, J. R., and Reichlin, S.**, Thyrotropin releasing hormone-diphtheria toxin related polypeptide conjugates: potential role of the hydrophobic domain in toxin entry, *J. Biol. Chem.*, 258, 1565, 1983.
105. **Thorpe, P. E. and Ross, W. C. J.**, The preparation and cytotoxic properties of antibody-toxin conjugates, *Immunol. Rev.*, 62, 119, 1982.
106. **Chakravarty, P. K., Carl., P. L., Weber, M. J., and Katzenellenbogen, J. A.**, Plasmin-activated prodrugs for cancer chemotherapy. II. Synthesis and biological activity of peptidyl derivatives of doxorubicin, *J. Med. Chem.*, 26, 638, 1983.
107. **Nisshin Flour Milling Co., Ltd.**, 1-α-D-arabinofuranosylcytosine derivative, *Jpn. Kokai Tokyo Koho, JP*, 58, 140, 099 [83, 140, 099], 19 Aug. 1983.
108. **Whiteley, J. M., Nimec, Z., and Galivan, J.**, Treatment of Reuber H35 hepatoma cells with carrier-bound methotrexate, *Mol. Pharmacol.*, 19, 505, 1981.
109. **Chu, B. C. F. and Howell, S. B.**, Differential toxicity of carrier-bound methotrexate toward human lymphocytes, marrow and tumor cells, *Biochem. Pharmacol.*, 30, 2545, 1981.
110. **Chu, B. C. F., Fan, C. C., and Howell, S. B.**, Activity of free and carrier-bound methotrexate against transport deficient and high dihydrofolate dehydrogenase-containing methotrexate-resistant 1210 cells, *J. Natl. Cancer Inst.*, 66, 121, 1981.
111. **Jaton, J. C. and Waron, H. U.**, Antibodies to folic acid and methotrexate obtained with conjugates of synthetic polypeptides, *Arch. Biochem. Biophys.*, 122, 157, 1967.
112. **Hurwitz, E., Levy, R., Maron, R., Wilchek, M., Arnon, R. and Sela, M.**, The covalent binding of daunomycin and adriamycin to antibodies with retention of both drug and antibody activities, *Cancer Res.*, 35, 1175, 1975.
113. **Hazarika, P., Elliot, P. G., and Byrom, N.**, The cytotoxic activity of chemically modified antitumor antibodies, in *Proc. 25th Colloq. Proteins of the Biological Fluids*, Brugge, Belgium, 1977, 156.
114. **Hsieh, K. and Marshall, G. R.**, Alkylating angiotensin II analogues: synthesis, analysis, and biological activity of angiotensin II analogues containing the nitrogen mustard melphalan in position 8, *J. Med. Chem.*, 24, 1304, 1981.
115. **Paiva, T. B., Paiva, C. M., Freer, R. J., and Stewart, J. M.**, Alkylating analogs of peptide hormones. II. Synthesis and properties of ρ-[*N,N-bis*(2-chloroethyl)amino]phenylbutyryl derivatives of angiotensin II, *J. Med. Chem.*, 15, 6, 1972.
116. **Freer, R. J. and Stewart, J. M.**, Alkylating analogs of peptide hormones. I. Synthesis and properties of ρ-[*N,N-bis*(2-chloroethyl)amino]phenylbutyryl derivatives of bradykinin and bradykinin potentiating factor, *J. Med. Chem.*, 15, 1, 1972.
117. **Robinson, D. A., Whiteley, J. M., and Harding, N. G. L.**, Cell-directed antimetabolities: alternative synthesis of cytotoxic methotrexate-containing macromolecules, *Biochem. Soc. Transcr.*, 1, 722, 1973.
118. **Harding N. G.**, Amethopterin linked covalently to water soluable macromolecules, *Ann. N.Y. Acad. Sci.*, 186, 270, 1971.
119. **Jaffe, B. M., Smith, J. W., Newton, W. T., and Parker, C. W.**, Radioimmunoassay for prostaglandins, *Science*, 171, 494, 1971.
120. **Erlanger, B. F., Borek, G., Beiser, S. M., and Lieberman, S.**, Steroid-protein conjugates. II. Preparation and characterization of conjugates of bovine serum albumin with progesterone deoxycorticosterone and esterone, *J. Biol. Chem.*, 234, 1090, 1959.
121. **Anderson, G. W., Zimmerman, J., and Callahan, F. M.**, A reinvestigation of the mixed carbonic anhydride method of peptide synthesis, *J. Am. Chem.*, 86, 1839, 1964.
122. **Butler, G. C., Harington, C. R., and Yuill, M. E.**, Observations on antisera against aspirine-protein complexes, *Biochem. J.*, 34, 838, 1940.
123. **Clutton, R. F., Harington, C. R., and Yuill, M. E.**, Studies in synthetic immunochemistry. III. Preparation and antigenic properties of thyroxyl derivatives of proteins and physiological effects of their antisera, *Biochem. J.*, 32, 1119, 1938.
124. **Lauer, R. C., Solomon, P., Hakanishi, K., and Erlanger, B. F.**, Preparation and specificity of antibodies to insect C16-juvenile hormone, *Fed. Proc., Fed. Am. Soc. Exp. Biol.*, 32, 500, 1972.
125. **Masquelier, M., Baurain, R., and Trouet, A.**, Amino acid and dipeptide derivatives of daunorubicin. I. Synthesis, physicochemical properties, and lysosomal digestion, *J. Med. Chem.*, 23, 1166, 1980.
126. **Thorpe, P. E. and Ross, W. C. J.**, The preparation and cytotoxic properties of antibody-toxin conjugates, *Immunol. Rev.*, 62, 119, 1982.
127. **Moolten, F. L. and Cooperband, S. R.**, Selective destruction of target cells by diphtheria toxin conjugated to antibody directed against antigens on the cells, *Science*, 169, 68, 1970.

128. **Carlsson, J., Drevin, D., and Axen, R.,** Protein thiolation and reversible protein-protein conjugation. *N*-succinimidyl-3-(2-pyridyldithio)propionate — a new heterobifunctional reagent, *Biochem. J.*, 173, 727, 1978.
129. **Bumol, T. F., Wang, Q. C., Reisfeld, R. A., and Kaplan, N. O.,** Monoclonal antibody and an antibody-toxin conjugate to a cell-surface proteoglycan of melanoma cells suppress in vivo tumor growth, *Proc. Natl. Acad. Sci. U.S.A.*, 80, 529, 1983.
130. **Distefano, P. S., Schweitzer, J. B., Taniuchi, M., and Johnson, E. M.,** Selective destruction of nerve growth factor receptor-bearing cells in vitro using a hybrid toxin composed of ricin A chain and a monoclonal antibody against the nerve growth factor receptor, *J. Cell Biol.*, 101, 1107, 1985.
131. **Upton, J. L., Woods, R., Jessen, G. L., de L. Castrucci, A. M., Hadley, M. E., Wilkes, B. C., and Hruby, V. J.,** Superpotent melanotropins and mouse prelage melanogenesis, in *Pigment Cell 1985*, Bagnara, J. T., Klaus, S. N., and Schartl, M., Eds., University of Tokyo Press, 1984, 139.
132. **Levine, N., Lemus-Wilson, A. M., Wood, S. H., Abdel-Malek, Z. A., Hruby, V. J., and Hadley, M. E.,** Follicular melanogenesis in the mouse: stimulation by injected and topical melanotropins, *J. Invest. Dermatol.*, 89, 270, 1987.
133. **Hadley, M. E., Wood, S. H., Lemus-Wilson, A. M., Dawson, B. V., Levine, N., Dorr, R. T., and Hruby, V. J.,** Topical application of a melanotropic peptide induces systemic hair follicular melanogenesis, *Life Sci.*, 40, 1889, 1987.
134. **Dawson, B. V., Hadley, M. E., Don, S., and Hruby, V. J.,** *In vitro* animal and human cadaver skin models for percutaneous delivery of melanotropin analogues, *Abst. 13th Intl. Pigment Cell Conf.*, 1986, 32.
135. **Richman, S. P., Woodcock, T. M., Kubota, T. T., Blumenreich, M. S., Gentile, P. S., and Allegra, J. C.,** Phase II trial of vinblastine, bleomycin, and cisplatin (VBP) followed by dacarbazine and mitolactol in metastatic melanoma, *Cancer Treat. Rep.*, 68, 1395, 1984.
136. **Landfield, P. W., Baskin, R. K., and Pitler, T. A.,** Brain aging correlates retardation by hormonal-pharmacological treatments, *Science*, 214, 581, 1981.
137. **DeWied, D. and van Ree, J. M.,** Neuropeptides, mental performance and aging, *Life Sci.*, 31, 709, 1982.
138. **Beckwith, B. E. and Sandman, C. A.,** Central nervous system and peripheral effects of ACTH, MSH and related neuropeptides, *Peptides*, 3, 411, 1982.
139. **Strand, F. L. and Kung, T. T.,** ACTH accelerates recovery of neuromuscular function following crushing of peripheral nerve, *Peptides*, 1, 135, 1980.
140. **Strand, F. L. and Smith, C. M.,** LPH, ACTH, MSH and motor systems, *Pharmacol. Ther.*, 11, 509, 1980.
141. **Murphy, M. T., Richards, D. B., and Lipton, J. M.,** Antipyretic potency of centrally administered α-melanocyte stimulating hormone, *Science*, 221, 192, 1983.

Chapter 10

THE MELANOTROPIC PEPTIDES: A SUMMARY

Aaron B. Lerner and Gisela E. Moellmann

The melanotropic peptides were discovered and named because of their vital importance in the control of pigmentation of frogs, tadpoles, chameleons, lizards, and fish. In dilutions of 10^{-10} *M* and higher, they bring about the dispersion of pigment granules in melanocytes to darken the integument of these animals. Melanotropic peptides can also darken the skin of human beings. The occurrence of these potent agents at extrapituitary sites such as the mammalian CNS, raises the question as to whether they affect structures other than pigment cells.

As described in earlier chapters, the amino acid sequences of the melanotropic peptides occur in the precursor molecule, pro-opiomelanocortin (POMC). Intracellular processing of POMC in the pituitary gland can lead to the release of α-, β-, and γ-melanocyte-stimulating hormone (MSH). Although a molecular relative of α- and β-MSH, γ-MSH is not known to have an effect on melanocytes, but may play a role in the activity of adrenocorticotropin (ACTH). β-MSH is probably not released under normal conditions, but remains part of β-lipotropin. α-MSH, which appears after the processing of POMC or ACTH, is present in many parts of the CNS and has been the subject of most of the recent studies on melanotropic peptides. α-MSH has been implicated as being important in a variety of nonmelanocytic functions, shown in Table 1. Thus far, except for the possibility of α-MSH counteracting the effect of interleukin-1, as for example in the production of fever, its actions are not clear-cut. We are used to thinking of hormones, such as insulin, ACTH, thyroxin, and others, as having specific, single dramatic effects. With the melanotropic peptides, which are now included in the class of neuropeptides, we may have to look at the subject differently. For example, perhaps three or more neuropeptides, including α-MSH, act synergistically, and all must be increased, whereas at the same time the levels of two other peptide hormones must be decreased to obtain an effect. Thus, one would need a change in the total profile of these peptides and not simply an elevation or decrease in one. To study this problem it would be necessary to assay, in human beings, at least ten peptides simultaneously. These agents would include POMC, α-, β-, and γ-MSH, deacetylated α-MSH, ACTH, acetylated ACTH, CLIP, β-lipotropin, enkephalin, and possibly others. Patterns of the level of these peptides in the blood of human beings could give us information as to whether or not these substances play a role in a variety of disorders.

Another question to be asked at this time is whether or not there are peptides other than those mentioned above that are processed from POMC and are biologically active. Looking at the peptides that are known to be processed from the original molecule, one is impressed that all are in the right-hand quadrant of the chart (Figure 1). What is happening on the left-hand side? Can we assume that small peptides are liberated from the left-hand side as well? If so, how can they be detected and assayed? Whenever one of us (A.B.L.) carried out isolation work, he realized that what was being discarded could be more important than the active agent being isolated. There could be numerous sequences of great importance still hidden from view in the parental POMC molecule.

The experimental modifications in peptide structure by Hruby and associates have demonstrated that hormonal activity can be increased dramatically. There are numerous ways of changing peptide molecules. The work with α-MSH has led the field.

Table 1
ACTIONS OF α-MSH IN MAMMALS OTHER THAN ON PIGMENTATION

Modulating the secretion of hormones
Interactions with neurotransmitters
Inhibiting immunostimulatory and inflammatory actions of Interleukin-1
Thermoregulation
Fetal growth
Motivation, learning, memory, and behavior
Sexual excitement
Premenstural syndrome

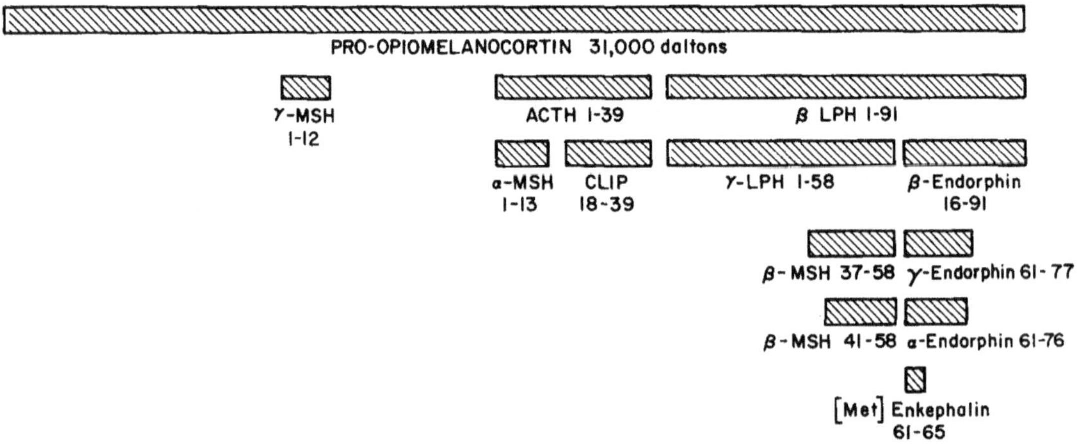

FIGURE 1. The distribution of biologically active peptides in pro-opiomelanocortin.

INDEX

A

Abrin, 138
Acetylated ACTH, 151, 152
Acetylated-[4-cysteine, 10-cysteine] α-MSH analogs, see also Analogs of α-MSH, cyclic
  conformations of, 97—99
  relative potencies, 85, 87, 88
Acetylated-[4-homocysteine, 10-cysteine] α-$MSH_{4-13}NH_2$, 86, 88
Acetylated-[4-norleucine, 7-D-phenylalanine] α-MSH, tritiated, 134
ACTH, 62, 151, 152
  in Addison's disease, 131
  analogs
    biomedially-active conjugates, 138, 142
    structural studies, spectroscopy, 96, 101—103
  daunomycin conjugates, 138
  mechanisms of action, 69
  MSH-ACTH structure-activity relationships, 69—71
  pigmentation and, 130
  polyphosphoinositide hydrolysis and, 29, 40
  primary structure, 70, 132
  protein phosphorylation, 38, 39, 41, 42
  sequence homologies, 48
  xanthophore dispersion, 40—41
ACTH-like peptides, frog skin bioassay, 4
ACTH receptors, 62
Actin, 20
Active ester method, 141
Activity, prolonged, 133
Addison's disease, 130, 131
Adenylate cyclase, see also Cyclic AMP
  assays
    melanoma cells, 10—11
    α-MSH activity, 61, 76
  α-MSH, 63—64, 68
  protein phosphorylation in melanophores and, 35
  receptor-calcium interactions, 120
  signal transduction, 122
Adrenal cells, protein phosphorylation in, 42
α-Adrenergic agonists, 123
β-Adrenergic agonists, 64
α-Adrenergic receptors, 17, 18, 20, 121, 123
Adriamycin, 137, 140, 142, 144
Affinity groups, α-MSH derivatives carrying, 114—115
Aggregation, melanosome, 20—22
Aging, 144
Alkaline phosphatase, 55
Amino acid sequences
  ACTH (1-29), 101
  ACTH and α-MSH, 69, 70, 132
  biomedically useful carrier synthesis, 139
  α-MSH, importance of, 80—84
    C-termini, 81—83
    D-phenylalanine substitution, 83, 84
    N-termini, 80—81
  MSH and MCH, 22
Analogs of α-MSH
  bioassays, 10, see also Bioassays
  biomedical applications, see also Biomedical applications
    agents with prolonged activity, 133
    resistance to inactivation, 133—134
    superpotent melanotropins, 132
  chemical structures, 69, 70
  comparative properties, 64
  linear
    C-terminal amino acids, 81
    dithol, 89
    NMR studies, 103—106
    N-terminal amino acids, 80, 81
    RP-HPLC, 95—96
  [4-norleucine, 7-D-phenylalanine]-substituted, 10, 12, 63, 64
  residual action, 12
  structure-activity relationships, 69—71, see also Structure, three-dimensional
Analogs of α-MSH, cyclic, 96
  biological properties of melanotropins, 76
  biomedical applications, 133
  carba analogs of acetyl-[4-cysteine, 10-cysteine] α-$MSH_{4-13}NH_2$, 88—89
  cyclized via 4 and 10 disulfide bridges, with cysteine substitution, 83—88
    ac-[4-D-cysteine, 10-cysteine] α-MSH_{4-13}-NH_2, 85—86, 88
    ac-[4-homocysteine,10 cysteine] α-MSH_{4-13}-NH_2, 86—88
    [4-mercaptoacetic acid, 10-cysteine] α-MSH_{4-13}-NH_2, 86, 88
    [4-mercaptopropionic acid, 10-cysteine] α-MSH_{4-13}-NH_2, 86, 88
    penicillamine substitution for Cys in 4 and 10 positions, 87, 88
  [4-cysteine,10-cysteine] α-MSH cyclized via cysteine disulfide bridges, 80—83
    C-terminal amino acids, 81—83
    D-phenylalanine substitution, 83, 84
    N-terminal amino acids, 80—81
  other biologic activities, 89
  RP-HPLC, 96
  structure-function studies, 77—80
    [Cys4,Cys10]-α-MSH, 78—80
    seven position, importance of, 77—78
Angiotensin, 142
Anolis bioassays, 4, 6, 7, 61
  α-MSH
    activity, 76
    analog properties, 64
    calcium requirements, 64, 65
    cAMP-mediated effects, 63

model of mechanism of action, 68
photoaffinity groups, 115
MSH regulation, heterospecific modulators, 70
receptors
calcium role, 119—121
irreversible stimulation of, 117—119
signal transduction, 122
Antagonists, 63, 64
Antipyretic activity, 144
Arachidonic acid, 68
Aspirin conjugates, 141
Assays
MSH activity, see Bioassays
α-MSH receptor, radioligands for, 62—63
β-MSH receptor, radioligands for, 51—57
ATP
cAMP-dependent protein phosphorylation, 28
α-MSH, model of mechanism of action, 68
protein kinases, see Protein phosphorylation
A23187, 64, 66, 69, see also Calcium
Autoinhibition, and melanosome movement, 23
Azide method, 141

B

BCNU, 137—138
Behavioral potencies, 89, 144
B-50 phosphorylation, 41
Bifunctional cross-linking method, 142—143
Binding, receptor, 60, 61, 63, 68
Bioassays, 2—12, see also *Anolis* bioassays; *Xenopus*; Melanoma cells
α-MSH activity, 76
Anolis (lizard) skin, 4, 6, 7
frog skin, 2—5
melanoma, 8—11
adenylate cyclase, 10—11
clonogenic, 11
in situ melanin, 11
tyrosinase, 8—10
melanophore index, 2, 3
α-MSH analog properties in, 64
prolonged (residual) melanotropin action, 12
teleost skin, 6, 8
Biocytin derivatives, 135, 136
Biodegradation, 4, 89, 133—134
Biomedical applications
analogs, 132—134
with prolonged activity, 133
resistance to enzymatic activation, 133—134
superpotent, 132—133
diagnostic probes, 134—136
drug delivery, 135—143
magic bullet concept, 136—137
prodrug approach, 138—139
transport and uptake efficiency, 139
transport vehicles, 137—138
drug synthesis strategies, 139—143
active ester method, 141
azide method, 141
bifunctional cross-linking, 142, 143
carbodiimide method, 140
mixed anhydride method, 140
oxidation-reduction method, 142
pigmentation, normal skin, 130
pigment disorders, 130—132
transdermal delivery of melanotropins, 142—143
Biostability, 4, 89, 133—134
Biotin-avidin system, 115
Biotin-melanotropin conjugates, 135, 136
Bleomycin, 144
Bradykinin conjugates, 142
Bullfrog (*Rana catesbiana*) skin bioassay, 2, 3, 5

C

Calcium
melanocyte-concentrating hormone and, 21
melanosome movement and, 18—19, 23
α-MSH receptor-mediated activity, 64—67
protein phosphorylation and, 29
lacrimal gland, 41
in melanophores, 35—36
receptor photoaffinity labeling, 119—121
signal transduction, 122
Calmodulin, 28, 29
Carba analogs, α-MSH, 88—89
Carbodiimide method, 140, 142
Carcinogenesis, 131
Carmustine, 137—138
Carotenoids, 40—41, 143
Catecholamines
melanosome movement and, 17, 18, 20
MSH regulation, heterospecific modulators, 70
receptor photoaffinity labeling, 121—123
CCNU, 137—138
Cell cycle, MSH responses, 52, 55
Central nervous system, protein phosphorylation in, 41—42
Central nervous system assays, α-MSH analogs, 89
Cesium, 18, 19
Chemical vitiligo, 131
Chlorambucil, 137, 140, 141
N-(2-Chloroethyl)-*N*-nitrosocarbamoyl group, 137
Cholera toxin, 122—123
Choline, 19
Chromatography, RP-HPLC, 94—96
Chromatophore index, 2
α-Chymotrypsin, 133—134
Circular dichroism studies, 96
Cisplatin, 144
CLIP, 151, 152
Clonidine, 121
Clonigenic bioassay, melanoma, 11
Cloudman melanoma cells, 61, see also Melanogenesis
bioassays, 8—11
cAMP-mediated α-MSH effects in, 63
genetic variants, 48—49
α-MSH analog properties in, 64
protein phosphorylation in, 36—40
Cognitive functions, 144

Colchicine, 19, 52
Competitive antagonists, 63, 64
Computer-assisted modeling studies, 96—99
Conjugates, MSH, see also Radioligands
  biomedical applications, 134—136
    biotin, 135, 136
    controlled delivery, 135—143
    drug and toxin, 138
    fluorescein, 135, 136
    synthetic strategies, 139—143
    transdermal delivery, 142—143
  macromolecular, 62, 67
Controlled drug delivery, 135—143
Cortisol, 131
Cross-linking, 131, 142—143
C-terminal amino acids
  biomedically useful carrier synthesis, 139
  α-MSH, importance of, 81—83
  α-MSH analogs
    activity of, 77, 81—83
    [4-cysteine, 10-cysteine] disulfide-linked, 78
  photoaffinity groups, 115
Cushing's disease, 131
Cyclic AMP
  in melanogenesis, 49
  melanosome movement and, 16—17, 23
  α-MSH
    model of mechanism of action, 68
    receptor-mediated activity, 63—64, 120
  MSH regulation, heterospecific modulators, 70
  protein phosphorylation and, 28—29
    lacrimal gland, 41
    in melanoma cells, 40
    in melanophores, 32, 35
    xanthophore dispersion, 40—41
  receptor interactions, 63—64, 120
  signal transduction, 122
Cyclic analogs, see Analogs of α-MSH, cyclic
D-Cysteine, α-MSH analogs, 85—86, 88
[4-Cysteine,10-cysteine] α-MSH analogs, see also Analogs of α-MSH, cyclic
  chemical structures, 69, 70
  comparative properties, 64
  conformations of, 69, 97—99
  design and preparation, 78—80
Cytochalasin B, 19
Cytoskeleton, and melanosome movement, 19—20, 23
Cytotoxicity, melanin precursors, 50

D

Dacarbazine, 144
D-amino acids
  ACTH analogs, CD and NMR spectroscopy, 96
  biomedical applications of analogs, 132—133
  melanoma bioassays, 11
  α-MSH, racemization in, 77—78
  α-MSH analogs, 85—86, 88
    activity of, 77
    biostability and, 89
    comparative properties, 64
    computer-assisted modeling studies, 97, 98
    HPLC, 95—96
    importance of, 81—83
  MSH-specific calcium requirements, 65, 67
  residual action and, 12
Daunomycin, 137, 138, 140—142, 144
Degradation
  analogs, 89
  assay considerations, 4
  biomedically-useful analogs, 133—134
Desacetylated α-MSH, 151
Desacetylated analogs, α-MSH, 38, 39, 77, see also Analogs of α-MSH
Diacylglycerol, 29, 40, 68
Diagnostic probes, 134—136
Dibenamine, 20
Dibutyryl CAMP, 17, 40
5,6-Dihydroxyindole, 50
Dispersion, melanosome, 16—20, 63
Disulfide bridges, cyclic analogs bridged by, 80—88
Dithiol analogs, α-MSH, 89
Dithiothreitol, 142
Dopachrome conversion factor, 50
Dopamine, 70, 122
Dopa oxidase, 50, 55
Dopaquinone, 50
Dose-response effects, MSH-specific calcium requirements, 65
Drug conjugates, see Biomedical applications; Conjugates, MSH
Dynein, 20
Dynorphin, 69

E

EC protein, 41
Eel, 7
Endorphins, 151, 152
  central nervous system, 41
  protein phosphorylation and, 38, 39
Enkephalins, 151, 152
Enzymatic inactivation, see Degradation
Epinephrine, 70
Erythrophores, 6
European minnow, 6, 8
Exocytosis-coupled protein, 41
Exopeptidases, 4

F

Fish, 61
  bioassays, 6, 8
  α-MSH regulation in, 69
  melanocyte concentrating hormone, 21
FITC-conjugated MSH, 116
Fluorescent conjugates
  biomedical applications, 135, 136
  MSH, 113—114, 116
Forskolin
  receptor interactions in *Anolis*, 120, 123

protein phosphorylation and, 34—36, 40
Free radicals, 50
Frog, see also *Rana pipiens*
bioassays
[4-cysteine, 10-cysteine] disulfide-linked analogs, 78, 80—82, 84, 85
in vitro, 2, 3, 5
in vivo, 4, 5
melanocyte concentrating hormone, 21
α-MSH regulation in, 63, 69, 70

G

Glucagon, 142
Glutamic acid residues, 139
Glutaraldehyde, 142
Grooming assay, 89
Growth regulation, 49

H

Heat-alkali treatment, α-MSH analogs, 77
Heterospecific regulation, α-MSH bioactivity, 69, 71
High-performance liquid chromatography (HPLC), reversed phase, 94—96
Homocysteine, cyclic analogs substituted with, 86—88
HPLC, see High-performance liquid chromatography
Human melanoma cells, cAMP-mediated α-MSH effects in, 63
Hyperpigmentary disorders, 131
Hypomelanotic disorders, 131

I

Indole-blocking factor, 50
Inositol-1,4,5-trisphosphate, 40, 42, 68
In situ melanin assay, 11
In vivo assay, frog skin, 4, 5
Iodine^{125} labeled MSH, 51—57, 113, 114
Ionotophoretic injection, 62
Irreversible melanosome dispersion, 64, 65
Irreversible receptor stimulation, 117—119
Isobutylmethylxanthine, 40, 41
Isoproterenol, 65

L

Lacrimal gland, 41
Lanthanum, 120
Learning and behavior functions, 144
Leopard frog, see Frog; *Rana pipiens*
β-Lipotropin, 41, 48, 151, 152
Lithium, 18, 19
Lizard, 61, see also *Anolis cardinensis*
Lomustine, 137—138
Lysine residues, 139
Lysosomal hydrolysis, 144

M

Macromolecular labels, MSH, 113—114
Magic bullet concept, 136—137
Magnesium, 120
Malignant melanoma, 131, see also Biomedical applications
therapeutic MSH conjugates, 138, 139
transdermal delivery systems, 142—143
Mason-Raper pathway, 50
MCH, see Melanocyte concentrating hormone
Mechanisms of action
melanogenesis, see Melanogenesis
melanosome movements, see Melanosome movements
α-MSH
molecular model, 67—69
MSH-ACTH, 69—71
peptide hormones, 60
Melanin precursors, cytotoxicity of, 50
Melanocyte concentrating hormone (MCH)
amino acid sequences, 22
melanosome movement and, 20—23
α-MSH regulation, 69
structure-activity studies, 22—23
tritiated, 112, 113
Melanocytes, α-MSH receptors, 61—62
Melanofilaments, 19—20
Melanogenesis, 16
cAMP role, 49
Cloudman melanoma genetic variants, 48—49
radiolabeled β-MSH for receptor analysis, 51—57
dopa phosphates, 53—57
receptor interactions, 51—53
regulators, 50
Melanoma, malignant, see Malignant melanoma
Melanoma cells
bioassays
adenylate cyclase, 10—11
clonigenic, 11
in situ melanin, 11
tyrosinase, 8—10
cAMP levels, MSH and, 17
α-MSH effects in, 63
MSH-receptor-binding studies, 116
Melanophore index, 2, 3
Melanosome dispersion, 16—20, 64, 65
Melanosome movements, 16—23
aggregation, 20—22
dispersion, 16—20
calcium requirement, 18—19, 64, 65
cAMP role, 16—17, 63
melanofilaments, 19—20
protein kinases, 17
receptors, separate, 17, 18
sulfhydryl requirement, 17—18
melanocytes, melanogenesis and melanosomes, 16
model of control, 23
structure-activity studies, 22
α-Melanotropin, 151, 152
in Addison's disease, 131, 132
amino acid sequence, 132
analogs, see Analogs
bioassays, see Bioassays
mechanism of action, see Protein phosphorylation

radioiodinated, 113, 114
receptors, see Receptors
structure, see Structure, three-dimensional
β-Melanotropin, 151, 152
bioassays, see Bioassays
mechanisms of action, see Melanogenesis
pigmentation and, 130
radioiodinated, 113, 114
γ-Melanotropin, 151, 152
Melanotropins
mechanisms of action, see Melanogenesis
melanocyte concentrating hormone and, 21
POMC and, 151, 152
Melatonin, 20, 23, 70
Melphalan, 140
Mercaptoacetic acid, 86, 88
Mercaptopropionic acid, 86, 88
Met-enkephalins, 151, 152
Methotrexate, 137, 139, 140
Methoxyverapamil, 120
Methylxanthines, 63
Microfilaments, 19—20
Microtrabecular lattice, 20
Microtubules, 19—20
Mitomycin C, 137
Mixed anhydride method, 140
Mutagenesis, 131
Myosin light-chain kinase, 29

N

Neuromuscular disorders, 144
Noradrenaline/norepinephrine
melanosome movement and, 20, 23
MSH regulation, heterospecific modulators, 70
receptor effects, 121, 122
[4-Norleucine,7-D-phenylalanine] α-MSH analogs, 77, 78
biomedical applications, 132—133
fluorescein and biotin conjugates, 135, 136
topical delivery, 142
MSH-specific calcium requirements, 65, 67
potency, 10, 12, 63, 64
Normetanephrine, 142
N-terminal amino acids, 69
analog activity, 80—81
biomedically-useful carrier synthesis, 139
photoaffinity groups, 115
Nuclear magnetic resonance, α-MSH, 99—106
analogs of, 100—106
peptide conformation, 100
Nuclear Overhauser enhancement, 100

O

Oxidation-reduction method, 142

P

Paranitrophenylalanine, 95
Pars intermedia, 130
Penicillamine, 87, 88
p57 protein, 40—41
Phaeomelanin-eumelanin shift, 142
Phenylalanine diastereomers
analogs, α-MSH
biostability and, 89
comparative properties, 64
HPLC, 95
importance of, 81—83
melanoma bioassays, 11
MSH-specific calcium requirements, 65, 67
potency, 10, 12, 63, 64
residual action and, 12
racemization, 78
Phenylglycine, 95
Phenylthiourea, 56
Phosphatases, 55
Phosphates, dopa, 53—57
Phosphatidylinositol 4,5-bisphosphate kinase (PIP2), 41—42, 68
Phosphatidylinositol 4-phosphate kinase (PIP kinase), 41—42
Phosphatidyl serine, 29
Phospholipase A, 68
Phospholipase C, 29, 68
Phospholipids
ACTH and, 40
calcium-phospholipid-dependent protein kinases, 28, 29, 41—42
α-MSH, 68
Phosphoproteins, see Protein kinases; Protein phosphorylation
Phosphotyrosine proteins, 49
Photoaffinity labeling
α-MSH derivatives carrying, 115—116
MSH-receptor characteristics, 117—123
Phoxinus laevis, 6, 8
Piebaldism, 131
Pigmentation, normal, 130
Pigment disorders, 130—132
Polylysine, 137, 139
Polyphosphoinositide hydrolysis, ACTH and, 40
Polysaccharide conjugates, MSH, 62
POMC (pro-opiomelanocortin), 41, 48, 151, 152
Prazosin, 121
Precursors, melanin, 50
Prolonged activity, 133
Propranolol, 17
Prostaglandins, 18, 64, 65, 140
Protein conjugates, MSH, 62
Protein kinase C, 29, 40
Protein kinases, 28—29, 40—42
assays, 61
in melanogenesis, 49
melanosome movement and, 17, 23
Protein phosphorylation, 28—42, 122, 123
calcium-dependent, 29
cAMP-dependent, 28—29
in melanogenesis, 49
melanoma cells, 36—40
melanophore systems, 29—36
calcium requirements, 35—36
cAMP in, 32, 35

protein phosphorylation in, 29—34
other α-MSH target tissues, 40—42
adrenal cells, 42
central nervous system, 41—42
lacrimal gland, 41
xanthophores, 40—41
Proteolysis, see Degradation

Q

Quin-2, 42

R

Racemization, α-MSH amino acids, 77—78
Radioligands, 113, 114
biomedical applications, 134—135
MSH and MCH derivatives, 112—114
β-MSH, 51—57
α-MSH receptor assay, 62—63
Rana berlandieri, 19
Rana pipiens, 61, see also Frog
bioassays, 2—5
cAMP-mediated α-MSH effects in, 63
melanosome movement, 19
α-MSH activity, 76
α-MSH regulation in, 69, 70
Receptors
characterization and visualization, 116—117
labeled MSH and MCH derivatives, 112—114
fluorescent and macromolecular labels for MSH, 113—114
radioiodinated, 113, 114
tritiated, 112, 113
melanocyte concentrating hormone, 21
melanosome dispersion, 17, 18
MSH derivatives with affinity groups, 114—116
photoaffinity labeling, 117—123
calcium, role of, 119—121
catecholamine effects, 121—122
irreversible stimulation of receptors, 117—119
signal transduction mechanisms, 122—123
radiolabeled β-MSH for receptor analysis, 51—57
dopa phosphates, 53—57
interactions of, 51—53
Receptors, signal transduction, 60—71
activity mediated by, 60—67
adenylate cyclase stimulation, 63—64
calcium requirements, 64—67
melanocutes, evidence for receptors on, 61—62
radioligands and radioreceptor assays, 62—63
hormone mechanisms of action, 60
mechanism of action, models of, 67—69
structure-activity and mechanisms of action relationships, 69—71
Regulation of α-MSH bioactivity, 69, 71
Residual action of melanotropins, 12
Retinoic acid-melanotropin conjugates, 143
Reversed phase high-performance liquid chromatography, 95—96
Ricin, 138
RP-HPLC, 94—96
Rubidium, 18, 19

S

Second messengers, see also Cyclic AMP; Signal transduction
mechanisms of action, 60, 61
protein kinases, types of, 28
Sequence homologies, 48
Signal transduction, 63, see also Receptors, signal transduction
adenylate cyclase bioassays, 11
calcium requirements, 64—67
mechanisms of action, model of, 60, 61
melanosome movement, 18
receptor photoaffinity labeling, 122—123
SKH2 mice, 56
Skin, 142—143
Skin bioassays, see Bioassays
SPDP, see N-Succinimidyl 3-(2-pyridyldithio)propionate
Spectroscopy, see Structure, three-dimensional
Stability, α-MSH analogs, 89
Structure
ACTH and α-MSH, 132
analogs, 69, 143
cyclic analogs, conformationally constrained, see Analogs of α-MSH, cyclic
primary, see Amino acid sequences
ACTH and α-MSH, 132
Structure-activity relationships, see also Structure, three-dimensional
ACTH, 69—71
analogs, linear and cyclic, 77—80
MCH and MSH, 22—23
Structure, three-dimensional
circular dichroism studies, 96
computer-assisted modeling studies, 96—99
history, 94
RPLC, 95—96
nuclear magnetic resonance studies, 99—106
N-Succinimidyl 3-(2-pyridyldithio)propionate (SPDP), 142
Sulfhydryl groups, melanosome dispersion, 17—18
Superpotent analogs, 132—133
Synbranchus marmoratus, 7, 69
Synchronized cultures, Cloudman melanoma cells, 52, 55
Synthesis, drug-MSH complexes, 139—143

T

Tadpole, 61, 64, 65
Teleosts, 61
bioassays, 6, 8
α-MSH regulation in, 69
melanocyte-concentrating hormone, 21
Tertiary structure, see Structure, three-dimensional
Theophylline, 17, 65
Thiol-substituted agents, 142

Thymidine, 52, 55
Thyroxine, 140
Tobacco mosaic virus conjugates, MSH, 62, 67
Toluene diisocyanate, 142
Topical delivery, 142—143
Toxin conjugates, 138, 142
Trabecular lattices, 20
Transdermal delivery, 142—143
Transport vehicles, biomedical, 137—138
Triggering, 60, 61, 68
Tritiated analogs, 112, 113, 134
Trypsin, 133—134
β-Tubulin, 32, 34
Tyrosinase, 50
  assays, 61
    melanoma, 8—10
    α-MSH activity, 76
  analog-induced, 133
  cAMP and, 36, 49, 63
  dopa phosphates and, 55
Tyrosine residues, 49, 95

U

Ultraviolet radiation, 131

V

Vasoactive intestinal peptide, 38, 39
Verapamil, 19
Vinblastine, 144
Virus conjugates, MSH, 62, 67
Visualization, MSH-receptor, 116
Vitiligo, 131

X

Xanthophores, 40—41
Xenopus
  receptors
    calcium role, 119—121
    irreversible stimulation of, 117—119
    signal transduction, 122
  skin bioassays, 115
Xenopus laevis, 61
  melanophore protein phosphorylation, 29—34
  α-MSH, calcium requirements, 64, 65
Xylene diisocyanate, 142

Y

Yohimbine, 121

For Product Safety Concerns and Information please contact our EU representative GPSR@taylorandfrancis.com Taylor & Francis Verlag GmbH, Kaufingerstraße 24, 80331 München, Germany

Batch number: 10731074

Printed by Printforce, the Netherlands